Electrochemical and Corrosion Behavior of Metallic Glasses

Vahid Hasannaeimi, Maryam Sadeghilaridjani, and Sundeep Mukherjee

Electrochemical and Corrosion Behavior of Metallic Glasses

MDPI • Basel • Beijing • Wuhan • Barcelona • Belgrade • Manchester • Tianjin • Tokyo • Cluj

AUTHORS

Vahid Hasannaeimi, Maryam Sadeghilaridjani and Sundeep Mukherjee
Department of Materials Science and Engineering,
University of North Texas,
Denton, Texas 76203, USA

EDITORIAL OFFICE

MDPI
St. Alban-Anlage 66
Basel, Switzerland

For citation purposes, cite as indicated below:

Hasannaeimi, V.; Sadeghilaridjani, M.; and Mukherjee, S. *Electrochemical and Corrosion Behavior of Metallic Glasses*; MDPI: Basel, Switzerland, 2021.

ISBN 978-3-03943-724-5 (Hbk)
ISBN 978-3-03943-723-8 (PDF)

doi:10.3390/books978-3-03943-723-8

Cover image courtesy of MDPI.

© 2021 by the author. The book is Open Access and distributed under the Creative Commons Attribution license (CC BY-NC-ND), which allows users to download, copy and build upon published work non-commercially, as long as the author and publisher are properly credited. If the material is transformed or built upon, the resulting work may not be distributed.

Contents

About the Authors — ix
Preface — xi

CHAPTER 1
Introduction to Corrosion — 1

1.1. Introduction — 1
1.2. Passivity — 2
1.3. Types of Corrosion — 2
1.3.1. Uniform Corrosion — 3
1.3.2. Pitting Corrosion — 3
1.3.3. Crevice Corrosion — 4
1.3.4. Galvanic Corrosion — 4
1.3.5. Stress Corrosion Cracking (SCC) — 5
1.3.6. Corrosion Fatigue — 5
1.3.7. Intergranular Corrosion — 5
1.3.8. Erosion Corrosion — 5
1.3.9. Selective Leaching (Dealloying) — 6
1.3.10. Cavitation — 6
1.3.11. Fretting Corrosion — 6
1.4. Thermodynamics and Kinetics of Corrosion — 7
1.4.1. Thermodynamics of Corrosion — 7
1.4.2. Kinetics of Corrosion — 9
1.4.2.1. Weight Loss — 9
1.4.2.2. Electrochemical Techniques — 9
1.4.2.3. Potentiodynamic Polarization — 9
1.4.2.4. Electrochemical Impedance Spectroscopy (EIS) — 10
1.5. Corrosion Prevention and Control — 12
1.5.1. Design Modification — 12
1.5.2. Inhibitors — 13
1.5.3. Coatings — 13
1.5.4. Material Selection — 13
References — 14

CHAPTER 2
Metallic Glasses and Rapid Solidification — 17

2.1. Metallic Glass Synthesis — 17
2.2. Processing and Microstructure Characterization — 18
2.3. Applications of Metallic Glasses — 20
2.4. Corrosion Mechanisms in Metallic Glasses — 20
References — 21

CHAPTER 3
Zirconium (Zr)-based Bulk Metallic Glasses and Their Composites — 25

3.1. Zr-based Bulk Metallic Glasses — 25
3.2. Corrosion Behavior of Zr-Based Metallic Glasses — 25
3.3. Effect of Alloying Elements — 26
3.3.1. Effect of Copper (Cu) Addition — 26
3.3.2. Effects of Niobium (Nb) and Cobalt (Co) Addition — 28
3.3.3. Effect of Silver (Ag) Addition — 28
3.3.4. Effect of Rare-Earth (RE) Elements Addition — 29
3.4. Combined Effects of Mechanical Loading and Corrosion — 29
3.5. Effects of Structure and Crystallinity — 29
3.6. Zr-Based Bulk Metallic Glasses Composites — 30
3.7. Effect of Test Environment — 31
References — 33

CHAPTER 4
High-Density Metallic Glasses — 43

4.1. Iron (Fe)-based Metallic Glasses — 43
4.1.1. Effect of Alloying Elements — 43
4.1.1.1. Effect of Chromium (Cr) Addition — 44
4.1.1.2. Effect of Molybdenum (Mo) Addition — 45
4.1.1.3. Effect of Other Metals — 46
4.1.1.4. Effect of Metalloid Addition — 46
4.1.2. Effects of Structure and Crystallinity — 47
4.1.3. Effect of Test Environment — 48
4.2. Ni-Based Metallic Glasses — 52
4.2.1. Effect of Alloying Elements — 52
4.2.2. Effects of Structure and Crystallinity — 53

4.2.3. Effects of Test Environment	54
4.3. Cobalt (Co)-Based Metallic Glasses	54
4.4. Copper (Cu)-Based Metallic Glasses	55
4.4.1. Effect of Alloying Elements	55
4.4.2. Effect of Test Environment	57
4.5. Chromium (Cr)-Based Metallic Glasses	57
References	58

CHAPTER 5
Low-Density Metallic Glasses — 67

5.1. Titanium (Ti)-Based Metallic Glasses and Composites	67
5.1.1. Effect of Alloying Elements	68
5.1.2. Effects of Structure and Crystallinity	68
5.2. Ti-Based Bulk Metallic Glass Composites	69
5.3. Magnesium (Mg)-Based Metallic Glasses	71
5.3.1. Effect of Alloying Elements	72
5.3.2. Effects of Structure and Crystallinity	73
5.3.3. Mg-Based Metallic Glass Composites	73
5.4. Calcium (Ca)-Based Bulk Metallic Glasses	74
5.5. Aluminum (Al)-Based Bulk Metallic Glasses	74
References	75

CHAPTER 6
Noble Metal- and Rare-Earth-Based Metallic Glasses — 81

6.1. Noble Metal-Based BMGs	81
6.2. Rare-Earth Elements-Based BMGs	82
References	83

CHAPTER 7
Concluding Remarks — 87

About the Authors

Vahid Hasannaeimi obtained his Ph.D. in Materials Science and Engineering from the University of North Texas (2019), where he studied electrochemical and catalytic behavior of metallic glasses. He is currently a Postdoctoral Research Associate in the Department of Materials Science and Engineering at UNT. He received his M.S. from Tarbiat Modares University (2011) in Materials Science and Engineering, where he worked on the development of corrosion-resistant nanocomposite coatings. His research interests include surface degradation mechanisms in materials, functional and structural coatings, and *in situ* corrosion mechanisms in multiphase alloys.

Maryam Sadeghilaridjani is currently a Postdoctoral Research Associate in the Department of Materials Science and Engineering at University of North Texas. Prior to joining UNT, she worked as a Postdoctoral Fellow at Tohoku University, Japan. She received her Ph.D. from Tohoku University (2016), M.S. (2009), and B.S. (2006) from the University of Tehran, Iran. M. Sadeghilaridjani's research interests include processing and characterization of metallic glasses and multi-principal alloys including small-scale mechanical and corrosion behavior.

Sundeep Mukherjee is currently an Associate Professor in the Department of Materials Science and Engineering at the University of North Texas. Prior to joining UNT, he worked as a Postdoctoral Research Associate at Yale University (2011–2012) and Senior Engineer at Intel Corporation (2005–2011). Prof Mukherjee received his B.S. (1998) from Indian Institute of Technology, M.S. (2003) and Ph.D. (2005) from California Institute of Technology. Prof Mukherjee has published more than 100 papers in reputed international journals and given many invited talks and keynote lectures at conferences and universities. He has organized symposiums in several international conferences and serves on the editorial board of three journals. His research interests include structure–property relationships in metallic glasses and concentrated alloys.

Preface

Metallic glasses are multi-component metallic alloys with disordered atomic distribution, unlike their crystalline counterparts with long range periodicity in the arrangement of atoms. This amorphous microstructure in metallic glasses leads to unique and exceptional properties including high strength and hardness, superior wear and corrosion resistance, and soft magnetism, to name a few. In addition, metallic glasses may be thermoplastically processed in the supercooled liquid region above their glass transition temperature and net-shaped into complex geometries in a wide range of length scales not achievable using conventional methods. There are numerous technical papers on synthesis, processing, and properties of metallic glasses. These include many different alloy systems, various synthesis routes, and characterization of their mechanical, physical, chemical, and magnetic properties. Metallic glasses of different compositions are being commercially used in bulk form and as coatings because of their excellent corrosion resistance. One may "simplistically" attribute this characteristic to their amorphous structure with no grains/grain-boundaries and chemical homogeneity down to the atomic scale. However, the corrosion behavior of amorphous alloys with slightly dissimilar chemistries has been reported to be vastly different, indicating that there is limited understanding of the underlying electrochemical mechanisms.

This book was written with the objective of providing a comprehensive overview of the electrochemical and corrosion behavior of metallic glasses in a wide range of compositions. Corrosion in structural materials leads to rapid deterioration in the performance of critical components and serious economic implications, including property damage and loss of human life. The discovery and development of metallic alloys with enhanced corrosion resistance will have a sizable impact in a number of areas including manufacturing, aerospace, oil and gas, nuclear industry, and load-bearing bio-implants. The corrosion resistance of many metallic glass systems is superior compared to conventionally used alloys in different environments. In this book, we discuss in detail the role of chemistry, processing conditions, environment, and surface state on the corrosion behavior of metallic glasses and compare their performance with conventional alloys. Several of these alloy systems consist of biocompatible and non-allergenic elements, making them attractive for bioimplants, stents, and surgical tools. To that end, critical insights are provided on the biocorrosion response of several metallic glass systems in simulated physiological environments.

The book begins with a short introduction on the theoretical concepts of corrosion and different types of corrosion. The thermodynamics and kinetics of electrochemical processes are discussed, followed by common techniques for measuring the corrosion rate. Finally, various strategies for corrosion prevention are presented, including design modification, use of inhibitors, material selection, and use of protective coatings.

The second chapter is related to the development of metallic glasses, their processing–microstructure relations, and structural and functional applications. Based on the importance and applications of different alloy systems, the present book is divided into Zirconium (Zr)-based metallic glasses and their composites; high-density metallic glasses, including Iron (Fe)-, Nickel (Ni)-, Cobalt (Co)-, Copper (Cu)-, and Chromium (Cr)-based metallic glasses; low-density metallic glasses, including Titanium (Ti)-, Magnesium (Mg)-, Calcium (Ca)-, and Aluminum (Al)-based metallic glasses; noble metal-based alloys; and rare-earth elements-based metallic glasses.

For each category of alloys, in the third, fourth, fifth, and sixth chapters, the effects of composition, microstructure, test environment, and processing conditions on the corrosion performance are discussed. The biocorrosion response of several biocompatible metallic glasses is discussed in simulated physiological environments and compared with conventional crystalline alloys. The last chapter summarizes the latest findings on the electrochemical characteristics of metallic glasses and identifies several open questions and key issues in the fundamental understanding of their corrosion behavior.

This work was partly supported by funding from the National Science Foundation (NSF) under Grant Numbers 1561886, 1919220, and 1762545. Any opinions, findings, and conclusions expressed in this book are those of the authors and do not necessarily reflect the views of the National Science Foundation (NSF).

1. Introduction to Corrosion

1.1. Introduction

Corrosion is broadly defined as the chemical or electrochemical reaction of a material with its environment, resulting in the degradation of its surface and bulk properties [1]. Metals are used in a wide range of load-bearing applications and remain irreplaceable in many important areas due to their high strength, toughness, and predictable failure. However, corrosion is a major limitation in the case of metals because they react with the environment of use, albeit at different rates and to varying degrees [2]. Corrosion mitigation remains a major priority in several industries to prevent catastrophic failures and accidents. About 258 natural gas accidents due to corrosion in pipelines were reported in 2004 [2] and the numbers have continued to rise. This indicates that corrosion in metals is not well understood and controlled due to the complexity of interactions with their environment and insufficiency of protection methods. Corrosion-related accidents lead to major economic losses and are a huge concern for the safety of personnel and property. The direct annual cost of corrosion across the globe was reported to be approximately 3 percent of global gross domestic product (GDP) [3]. In the United States, it is estimated that US$ 2–4 trillion is lost due to corrosion-related failures each decade. However, the true cost of corrosion damage is likely much more and difficult to estimate. Therefore, the study of corrosion and its different forms is necessary to design and choose suitable materials for specific applications and maintain safety standards [4].

Corrosion may be classified in different ways such as low- and high-temperature corrosion or wet and dry corrosion. Wet corrosion occurs in aqueous solutions whereas dry corrosion typically occurs at high temperatures and in the absence of a liquid medium [5]. Corrosion in aqueous environments takes place by the electrochemical mechanism through half-cell reactions at the interface between the metal surface and an electrolyte, namely anodic (oxidation) and cathodic (reduction) reactions [6]. At anodic sites, an oxidation reaction occurs, which is the loss of electrons as:

$$A \rightarrow A^{n+} + ne^- \tag{1-1}$$

Simultaneously, a reduction process takes place at cathodic sites, which is the gain of electrons as [1]:

$$B^{n+} + ne^- \rightarrow B \tag{1-2}$$

Anodic and cathodic reactions proceed at the same rate due to charge balance. The anodic and cathodic sites may be formed on the surface of a metal due to some heterogeneity (such as composition or grain size differences, surface roughness, impurities or inclusions, and localized stresses) or between two dissimilar metals exposed to the corrosive environment. In general, there are four essential requirements for electrochemical corrosion: (1) an anodic reaction, (2) a cathodic reaction, (3) presence of an electrolyte, and (4) electrical connectivity [7].

1.2. Passivity

A material in a particular environment will generally show three types of corrosion behavior: active, passive, and immune. Active behavior refers to the dissolution of a material in its environment; passive indicates the formation of a protective surface layer which slows down the corrosion rate, while immunity refers to a lack of driving force for anodic dissolution in the environment. Passivity is the formation of a protective film on the metal surface due to chemical or electrochemical activity. The surface layer must be adherent and dense in order to be protective. Materials that are likely to form a passive layer on their surface are less affected by the environment. The passive film thickness in transition metals (e.g., Fe, Cr, Co, Ni, Mo) and their alloys (e.g., the Fe–Cr stainless steels) are tens to hundreds of angstroms (Å) [2]. The protective film on titanium has been reported to be about 250 Å in thickness after 4 years of exposure to ambient air [2]. Passive films are thicker in non-transition metals (e.g., zinc (Zn), cadmium (Cd), Cu, Mg, lead (Pb)) in the range of thousands to tens of thousands of angstroms. The thickness of oxide film naturally formed under ambient conditions on aluminum is 30–40 Å [2], while much thicker passive films (~4000 Å) may be formed by anodizing techniques [2]. Special surface analytical techniques are necessary to study the nature of these films, including X-ray photoelectron spectroscopy (XPS), scattering ion mass spectrometry (SIMS), Mössbauer spectroscopy, and X-ray absorption spectroscopy (XAS) [2].

1.3. Types of Corrosion

Corrosion may be classified based on the underlying mechanism and proceeds at different rates. There are two major types of corrosion: uniform corrosion and localized corrosion [2], as shown in Figure 1.1. In uniform corrosion (Figure 1.1a), a chemical reaction takes place evenly across the whole surface. Localized corrosion has various forms and may occur in specific areas in a material when exposed to an electrolyte (Figure 1.1b–k) [8].

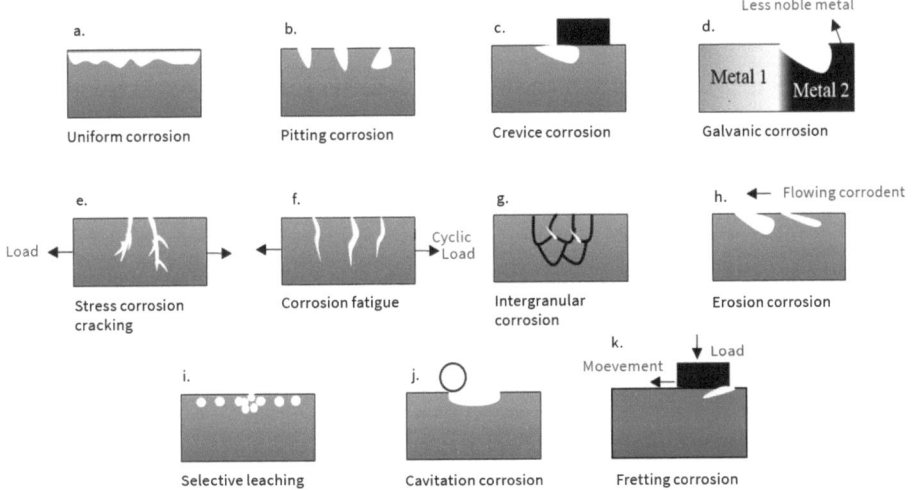

Figure 1.1. Different forms of corrosion. Source: Image by the authors.

1.3.1. Uniform Corrosion

Uniform corrosion is evenly distributed over the entire surface of the metal (Figure 1.1a). There is no preferential attack but a relatively uniform thickness reduction until the material ultimately fails. Homogeneous materials that do not form a passive layer in the corrosive environment are likely to undergo uniform corrosion. Since it affects a fairly large area of the metal, uniform corrosion is easier to detect and not considered to be dangerous as it is relatively more predictable [2,9].

1.3.2. Pitting Corrosion

Pitting corrosion is the most common form of localized corrosion (Figure 1.1b). A pit is initiated in a very small area on the metal surface and the mechanism may consist of: (1) local breakdown of the passive film (pit nucleation), (2) early pit growth, (3) stable growth of the pit, and (4) possible re-passivation [2,9]. Passive metals are typically vulnerable to pitting corrosion. Pits are nucleated at some chemical or physical heterogeneity in the passive film such as second phase particles, inclusions, mechanical damage, and solute-segregated grain boundary or dislocation [8]. The broken surface film acts as an anode, while the unbroken part acts as a cathode. The localized nature of the attack makes it difficult to detect and predict and may lead to sudden catastrophic failure of a component.

Pitting potential is the lowest potential where pitting corrosion may be initiated. The pitting potential decreases (becomes less noble) with increasing temperature, decreasing pH, and increasing ion concentration in chloride (Cl^-) or bromide (Br^-)

environments [8,10,11]. With the increase in temperature, the number of local defects in the passive film increases, leading to a decrease in the pitting potential [8,12,13]. The Cl$^-$ ion attack and dissolution rate of the protective layer may be limited with increasing pH due to the formation of an anion-selective diffuse barrier, which may lead to a thicker passive film and better corrosion behavior [14]. Electrolyte velocity is another important parameter that has a complex effect on pitting corrosion; 304 and 316 stainless steels undergo lesser pitting corrosion with increasing turbulence of solution. Aggressive ions are more likely to be swept away from the surface at higher solution velocity, which suppresses pit formation and thus, pitting potential shifts to the noble direction. Furthermore, at higher solution flow rates, the nucleated pit may be re-passivated and pitting potential may increase [8,15].

1.3.3. Crevice Corrosion

Crevice corrosion is a form of localized corrosion that takes place in crevices that are wide enough for liquids to penetrate but sufficiently narrow that the liquid cannot flow easily (Figure 1.1c). This form of attack may be underneath seals, bolt heads, in overlap joints and between tubes, inside screw threads, and strands of wires. Crevice corrosion may also occur beneath deposits, including corrosion products and dust particles [2]. This type of corrosion will occur due to the difference in the constituent's solution concentration, mainly oxygen. A metal within a narrow gap has less dissolved oxygen concentration acting as an anode, while the metal outside the gap, which is exposed to the bulk electrolyte, has a higher concentration of dissolved oxygen and acts as the cathode [2]. Materials that are passive or easily passivated such as stainless steels or aluminum undergo crevice corrosion in harsh environments containing chlorides or other salt solutions.

1.3.4. Galvanic Corrosion

Galvanic corrosion is an electrochemical process that occurs when two or more dissimilar metals are in contact in an electrolyte, in which one metal corrodes preferentially compared with others, as displayed in Figure 1.1d [2]. Dissimilar metals and alloys have different electrode potentials. The metal with the more negative electrode potential acts as the anode and undergoes corrosion, while the one with more positive potential acts as the cathode and remains un-attacked. The greater the potential difference, the higher the rate of galvanic corrosion. Galvanic corrosion is one of the most common forms of corrosion. Some examples of galvanic corrosion are: (1) steel pipes with brass fittings, (2) copper piping joined to steel tanks, (3) nickel alloy hull and steel rivets used in boats, and (4) zinc-coated screws in a copper sheet [2]. In certain cases, galvanic corrosion may be used for corrosion mitigation, known as cathodic protection [2].

1.3.5. Stress Corrosion Cracking (SCC)

Stress corrosion cracking is a type of corrosion that occurs by crack initiation and propagation in a metal from the combined action of applied stress and a chemical environment (Figure 1.1e). The stresses may be internal (e.g., residual stress) or external (i.e., applied stress). It may result in the sudden failure of metals/alloys and takes place in structures under stress such as pressure vessels, bridges and support cables, aircraft, pipelines, and turbine blades [2].

1.3.6. Corrosion Fatigue

Crack formation under the combined effect of repeated cyclic stress and a corrosive environment is known as corrosion fatigue as shown in Figure 1.1f. The mechanism for corrosion fatigue includes (1) pit nucleation in certain locations with high stress concentrations, (2) acceleration of both corrosion and mechanical degradation, and (3) hydrogen absorption from the environment resulting in embrittlement. Corrosion fatigue may occur in vibrating structures such as bridges and aircraft wings. Furthermore, orthopedic implants for knee and hip replacements in the human body may also experience corrosion fatigue under cyclic stresses [2].

1.3.7. Intergranular Corrosion

Corrosion taking place at or near grain boundaries is referred to as intergranular corrosion, as depicted in Figure 1.1g. Grains that undergo corrosion fail to resist stresses due to the weakening of cohesive forces between them and these result in a catastrophic reduction in mechanical strength and toughness. Impurity or elemental segregation at the grain boundaries typically results in intergranular corrosion. The 304 grade of stainless steel is vulnerable to intergranular corrosion when heated up to the temperature range of 425–790 °C, which is known as *sensitization*. During sensitization, carbon diffuses to the grain boundaries and combines with chromium to form chromium carbide precipitates. This decreases the chromium content locally from the areas in and adjacent to the grain boundaries to less than 12 at. % Cr required for passivation. Localized intergranular corrosion also occurs in certain aqueous environments [1,2].

1.3.8. Erosion Corrosion

Erosion corrosion is initiated due to the rapid movement of a corrosive liquid against a metal that attacks the surface. It consists of both mechanical and electrochemical processes (Figure 1.1h) [9]. Erosion corrosion may damage the passive film formed on the surface of a metal as well as the base metal. The mechanism for erosion corrosion includes: (1) surface impingement by the flowing liquid, (2) increased turbulence, and (3) wearing of the surface due to moving parts [9]. Erosion corrosion may occur in pumps, turbine parts, propellers, valves, heat exchanger tubes, bends, nozzles, and equipment exposed to fast-moving liquids [9]. Suspended solids in a liquid may also be a source of erosion corrosion as in coal slurry pipelines [2].

1.3.9. Selective Leaching (Dealloying)

The preferential removal of one or more elements from a solid solution alloy in the presence of an electrolyte as a result of an electrochemical oxidation reduction process is called selective leaching or dealloying, as illustrated in Figure 1.1i. Alloys composed of elements with a large difference between their electrode potentials are susceptible to selective leaching. The less noble elements dissolve in the electrolyte, resulting in a porous structure on the alloy surface. Selective leaching may be either uniform or localized. Some common examples of selective leaching include preferential removal of zinc from brass (dezincification) and iron from gray cast iron (graphitic corrosion) [2,5]. Decarburization (selective leaching of carbon), dealuminumification (selective removal of aluminum), decobaltification (selective removal of cobalt), and denickelfication (selective leaching of nickel) are other kinds of dealloying.

1.3.10. Cavitation

Cavitation corrosion results from the collapsing of gas bubbles formed in low-pressure zones when they rapidly enter a high-pressure zone and impact against the metal surface (Figure 1.1j). This form of attack occurs at high-flow velocities and under dynamic conditions near a metal surface and induces high local stresses and plastic deformation. Cracks are initiated and subsequently, particles from the material are removed. This form of corrosion may occur in water turbines, ship propellers, pump rotors, on hydrofoils, and on other surfaces subject to high-velocity flowing liquid, where the pressure suddenly changes [2].

1.3.11. Fretting Corrosion

Fretting corrosion (Figure 1.1k) involves the deterioration of contacting metals subjected to load when sliding against each other. Here, mechanical load plays a critical role in addition to the electrochemical processes. The mechanism of fretting corrosion is attributed to fine particles or fragments being released from one or both materials by adhesive wear. These fragments oxidize, forming hard debris which wear and destroy the metal surface. Fretting corrosion may also occur due to oxidation of the metal surface with relative motion between the parts. The oxide layer on the metal surface is removed by rubbing two surfaces against each other, exposing fresh and active metal surfaces, which oxidize again. This form of corrosion is typically seen in joints, press fits, vibrating equipment, joining rods, or springs. Fretting corrosion may also be active in orthopedic implants in the human body [2].

1.4. Thermodynamics and Kinetics of Corrosion

The thermodynamics and kinetics of the corrosion process determine whether a metal used in a given environment will corrode and the rate at which this takes place.

1.4.1. Thermodynamics of Corrosion

Thermodynamics predicts the tendency of a metal to corrode in a particular environment and the direction of an electrochemical reaction [8,16]. The Gibbs free energy change (ΔG) associated with an electrochemical reaction is calculated as [5]:

$$\Delta G = -nFE \qquad (1\text{-}3)$$

where n is the number of involved electrons in the anodic–cathodic reaction, F is Faraday's constant (=96,500 Coulomb/mole), and E is the electrode potential. Under standard conditions, the standard free energy of the reaction (ΔG^0) is related to the standard electrode potential (E^0):

$$\Delta G^0 = -nFE^0 \qquad (1\text{-}4)$$

Gibbs free energy change is the driving force for an electrochemical reaction. The thermodynamics of corrosion may be evaluated using potential–pH plots (Pourbaix diagrams), which are available for most common metals [17]. Such diagrams are constructed based on the Nernst equation:

$$\Delta G = \Delta G^0 + RT \ln Q_{reaction} \qquad (1\text{-}5)$$

where $Q_{reaction}$ is reaction activity, R is the gas constant, and T is the temperature. The typical reaction for a galvanic cell is:

$$bB + cC + \cdots \rightarrow pP + qQ + \cdots \qquad (1\text{-}6)$$

The above equation means that p moles of the component P and q moles of the component Q, etc., are formed as a result of the reaction of b moles of the component B plus c moles of component C, etc. $Q_{reaction}$ in Equation (1-5) is given by:

$$Q_{reaction} = \frac{a_P^p \times a_Q^q \cdots}{a_B^b \times a_C^c \cdots} \qquad (1\text{-}7)$$

where a_P^p is the activity of p moles of substance P [1].

Pourbaix diagrams are valuable in evaluating the activity of a metal or alloy in a corrosive environment. These diagrams show the regions of relative stability of the different phases in the system [17]. Figure 1.2 depicts the Pourbaix diagram for aluminum, showing the regions where Al (solid), Al_2O_3 or $Al(OH)_3$ (solid), Al^{3+} ions, and $[AlO_2]^-$ or $[Al(OH)_4]^-$ ions are stable [17,18]. The region on the Pourbaix diagram where the stable species is a dissolved ion is labeled as a region where corrosion occurs (Regions 2 and 4). The region of passivity is where the stable species is either a solid oxide or a solid hydroxide, in which the metal is protected by a surface film (Region 3). The region where the metal is unreacted is considered as the region of immunity (Region 1). The Pourbaix diagram for aluminum shows that corrosion may take place in highly acidic and highly alkaline conditions and a protection film may be formed at pH between 4 and 9 [2].

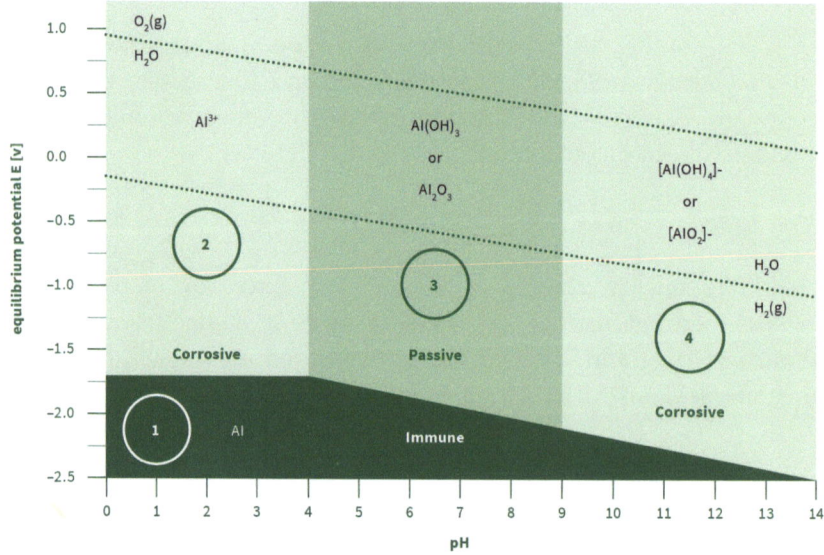

Figure 1.2. Pourbaix diagram of aluminum (Al) showing four different regions as Region (1) indicating immunity, Region (2) indicating acidic corrosion, Region (3) indicating passivation, and Region (4) indicating alkaline corrosion (redrawn from references [17,18]).

The application of Pourbaix diagrams in corrosion studies may be related to: (1) understanding of a metal's immunity to uniform corrosion in an electrolyte, (2) determining the passivity region of a metal over a wide range of pH and potential, (3) assessing the possible use of oxidizing inhibitors of corrosion, and (4) studying of localized corrosion cells [2,17]. Despite their advantages, Pourbaix diagrams have several limitations including: (1) assumption of equilibrium condition, (2) not useful for determining corrosion rates, (3) used mostly in the context of metals and not alloys, (4) all oxides or hydroxides are considered as passivity promoters, (5) do not consider localized corrosion, and (6) typically, these diagrams are available only for room temperature [1,2].

1.4.2. Kinetics of Corrosion

Pourbaix diagrams are useful for predicting the tendency for corrosion but do not give any information about the corrosion rates. Corrosion rates are measured typically by weight loss and/or electrochemical techniques.

1.4.2.1. Weight Loss

A simple direct experiment for quantifying corrosion rate is by weight loss measurement. The exposure of a clean weighed piece of material to the corrosive environment for a specified time leads to a reduction in its weight. The corrosion rate (CR) is then calculated as [19,20]:

$$CR = \frac{k\Delta w}{\rho At} \tag{1-8}$$

where k is a constant, Δw is the change in weight, ρ is density, A is the sample exposure area, and t is the exposure time.

1.4.2.2. Electrochemical Techniques

Electrochemical techniques are widely used for the study of corrosion behavior of materials in a variety of media. The primary electrochemical techniques that have been developed to measure the corrosion behavior of metals/alloys in an accelerated way include potentiodynamic polarization (Tafel analysis) and electrochemical impedance spectroscopy (EIS).

1.4.2.3. Potentiodynamic Polarization

Potentiodynamic polarization is a well-known electrochemical technique to study the corrosion mechanism and corrosion rate of materials using a potentiostat. Typically, three electrodes are used in this method, including (1) the sample as the working electrode, (2) a counter (auxiliary) electrode (such as platinum (Pt)) to provide the applied current, and (3) another electrode as a reference electrode (such as saturated Calomel or Ag/AgCl) to measure the potential. The corrosion potential and corresponding logarithm of current/current density are plotted, as shown schematically in Figure 1.3. The current represents the cathodic (region A in Figure 1.3) or anodic (regions B–D in Figure 1.3) reactions on the working electrode. The corrosion rate may be calculated using Faraday's law as [9]:

$$Q = \frac{nFW}{M} \tag{1-9}$$

where Q is the electrical charge (coulomb), F is Faraday's constant (96,500 coulombs/mole), n is the number of electrons involved in the electrochemical reaction, W is sample weight (g), and M is molecular weight (g). Based on Equation (1-9), the CR is calculated as:

$$W = \frac{QM}{nF} = \frac{Q(E.W.)}{nF} = \frac{i \times t(E.W.)}{F} \tag{1-10}$$

$$CR \ (mpy) = \frac{k \times i_{corr}(E.W.)}{\rho} \tag{1-11}$$

where mpy is the unit of corrosion rate in milli-inches (mils) per year, k is constant, i_{corr} is the corrosion current density ($\mu A.cm^{-2}$), E.W. is the equivalent weight (g), and ρ is the density ($g.cm^{-3}$). The corrosion current density (i_{corr}) may be determined from Tafel analysis by drawing tangents to the anodic and cathodic curves. The intersection of cathode and anode Tafel slopes determines the i_{corr} and the corrosion potential (E_{corr}), as shown in Figure 1.3. There are a number of important features on the curve in Figure 1.3. Region A involves a cathodic reaction and region B is the active region. Region C is known as the passive region in which the current density decreases with increasing potential until a low passive current density (i_{pass}) is reached, as shown by point I. The potential at point I refers to the passivation potential ($E_{passive}$). At point II, the applied current rapidly increases due to pitting (localized breakdown of passivity) or transpassive dissolution. This region is called the transpassive region (region D) [5]. A given system may contain some, but not necessarily all, of the features in the polarization scan shown in Figure 1.3.

1.4.2.4. Electrochemical Impedance Spectroscopy (EIS)

EIS represents a non-destructive technique to determine the corrosion performance of a wide range of materials within short periods. From EIS, the corrosion rate, electrochemical mechanism, and effectiveness of corrosion control methods may be evaluated [21]. Three electrodes (i.e., counter, reference, and working electrodes) are used in the EIS method similar to potentiodynamic polarization tests.

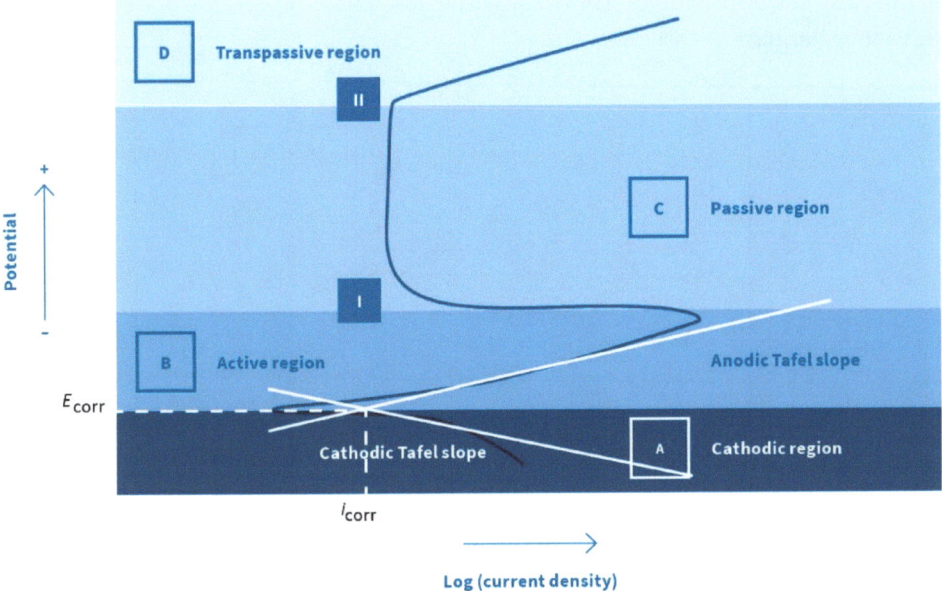

Figure 1.3. Schematic of polarization curve showing cathodic (region A), active (region B), passive (region C), and transpassive (region D) regions. Point I represents passivation potential, while point II represents pitting potential.

In EIS, the impedance of the electrochemical system is recorded as a function of frequency. A small amplitude alternating current (AC) signal is applied to the system to probe the impedance characteristics of a cell. Impedance (Z) determines the amplitude of current for a given voltage and is analyzed as a function of frequency (ω) as [1,22]:

$$Z(\omega) = \frac{E(\omega)}{I(\omega)} \qquad (1\text{-}12)$$

where E is the voltage ($E = E^0 \sin(\omega t)$), I is the current density ($I = I^0 \sin(\omega t + \varphi)$), t is time, and φ is phase shift. Z is described by the real component (Z′) and the imaginary component (Z″) as:

$$|Z|^2 = (Z')^2 + (Z'')^2 \qquad (1\text{-}13)$$

The most commonly used electrochemical impedance plot is known as the Nyquist plot, with -Z″ plotted versus Z′ (Figure 1.4). The impedance curves are used to obtain the electrical circuit elements, shown in the inset of Figure 1.4; R_s is the solution resistance, C_{dl} is the double layer capacitance, and R_p is the polarization resistance. Figure 1.4 shows a single capacitive semicircle, the radius of which is used to interpret the corrosion behavior. A wider semicircle represents better corrosion resistance (larger R_p) and vice versa.

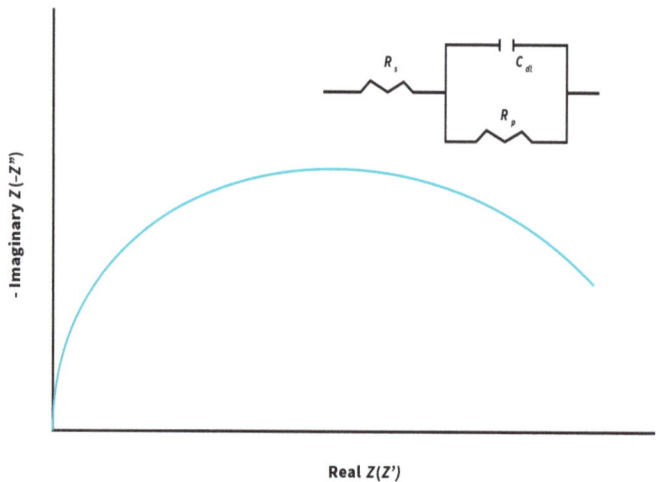

Figure 1.4. Nyquist plot obtained from the electrochemical impedance spectroscopy (EIS) test with -Z" plotted versus Z'.

1.5. Corrosion Prevention and Control

Corrosion prevention/protection may be accomplished by several methods such as design modification, inhibitors, coatings (cathodic protection or noble metal), and material selection [2]. The selection of one strategy or a combination thereof depends on the type of corrosion, the cost, nature of the material, and the working environment [2].

1.5.1. Design Modification

An appropriate design of a component may be used to prevent/retard corrosion. For instance, it is important to minimize areas where water may be trapped and accumulate leading to localized corrosion. Drilling holes in structures for proper drainage may help to address this issue. Another design strategy is not to use wood, paper, cardboard, open-cell foams, and sponge rubbers in high humidity environments as they absorb moisture and trigger corrosion in metallic parts. One way to prevent galvanic corrosion in structures is to insulate dissimilar metals/alloys by using coatings at their interface. Removing sharp corners may be beneficial in reducing crevice corrosion. It is critical to minimize unapproachable areas so that they are easily accessible for cleaning and maintenance [23].

1.5.2. Inhibitors

A corrosion inhibitor is a chemical (organic or inorganic) which is added to a solution, typically in small concentrations, to increase the corrosion resistance of a metal exposed to a corrosive environment [24]. The mechanisms for decreasing corrosion rate using inhibitors include altering the pH, retarding the cathodic and/or anodic processes, and formation of a passive layer [25]. Inhibitors may increase the pitting potential and decrease the number of corrosion pits per unit area as well as their growth rate. Inhibitors generally increase the time for crack initiation while decreasing the crack growth rate in stress corrosion cracking [2].

1.5.3. Coatings

(i) Cathodic protection: cathodic protection is a method used to prevent the corrosion of a metal surface by making it relatively cathodic in an electrochemical cell. By appropriate selection of metal pairs such as the use of an outer layer of zinc on steel substrate (galvanized steel), coatings may be employed as a method for corrosion protection. Since zinc has a more negative electrode potential than steel, it acts as an anode and steel is protected by the sacrificial dissolution of zinc [2].

(ii) Noble metal coatings: coating of nickel on steel is an example of noble metal coatings, since the electrode potential of nickel is more positive than that of steel. However, when defects are initiated in the noble metal coating, the steel substrate may undergo localized corrosion [2].

1.5.4. Material Selection

One of the most effective approaches for corrosion protection in a particular environment is choosing metals/alloys which are either unreactive or form a passive layer [2]. For instance, addition of more than 12 at. % Cr increases the corrosion resistance of steel due to the formation of a stable passive film. Titanium and its alloys and molybdenum-containing alloys are more corrosion resistant than conventional stainless steels. The presence of nitrogen in alloys which contain molybdenum is also favorable to corrosion resistance [2].

Some strategies for minimizing specific types of corrosion include:

(i) *Galvanic corrosion*: This may be avoided using metals with less electrode potential differences, using an insulator or coatings to separate the parts in galvanic pairs, avoiding the area effect caused by a small anode attached to a large cathode, using a third metal which is anodic with respect to the two metals in contact, and removing moisture between metal contacts [2,5].

(ii) *Crevice corrosion*: Crevice corrosion may be controlled by several ways including appropriate material selection, avoiding crevices, and cathodic protection [9].

(iii) *Pitting corrosion*: Methods to prevent pitting corrosion include material selection (e.g., Al-Mg, Ni-, Ti-based alloys, and sufficient addition of Cr, Ni, and Mo to stainless steel), cathodic protection, and changing the environment (e.g., pH, temperature, or concentration of species in the electrolyte) [9].

(iv) *Fretting corrosion*: Strategies to decrease fretting corrosion include using appropriate coatings on soft materials, applying lubricants, gaskets or sealing materials to impede the access of oxygen, decreasing adhesive wear, and finally, material selection (e.g., using hard materials) [9].

(v) *Stress corrosion cracking (SCC)*: This may be controlled by decreasing the stress below the threshold stress for fracture, changing the environment to be less corrosive, cathodic protection, using inhibitors, and appropriate addition of alloying elements (e.g., Ni addition to Fe-18 Cr stainless steels or Mo additions to Fe–Cr–Ni alloys), shot peening of the metal surface to induce a state of compressive stress (rather than tensile stress) on the surface, and annealing to remove residual stresses [2,5,9].

Proper material selection is one of the most effective strategies to reduce the corrosion rate. There have been notable advances with respect to the development of corrosion-resistant alloys (CRAs) over the past century. Metallic glasses, also known as amorphous metallic alloys, hold tremendous potential as CRAs. Since the first discovery in 1960 [26], a series of metallic glasses based on copper, palladium, zirconium, titanium, iron, and magnesium have been successfully fabricated [27]. Iron- and nickel-based metallic glasses have been developed with corrosion resistance better than conventional Ni-based superalloys [28,29]. More recently, some amorphous alloys have shown good biocompatibility in simulated body fluids in addition to excellent mechanical properties. To extend the range of potential applications for metallic glasses, the study of their corrosion behavior in different environments is of utmost importance.

References

1. Revie, R.W.; Uhlig, H.H. *Corrosion and Corrosion Control an Introduction to Corrosion Science and Engineering*, 4th ed.; John Wiley & Sons Inc.: Hoboken, NJ, USA, 2008.
2. McCafferty, E. *Introduction to Corrosion Science*; Springer: Alexandria, VA, USA, 2009.
3. Koch, G.H.; Brongers, M.P.H.; Thompson, N.G.; Virmani, Y.P.; Payer, J.H. *Corrosion Cost and Preventive Strategies in the United State*; National Technical Information Service Report No. FHWA-RD-01-156; Federal Highway Administration: Washington, DC, USA, 2001.
4. Hammonds, P. An Introduction to Corrosion and its Prevention. In *Comprehensive Chemical Kinetics*; Compton, R.G., Ed.; Elsevier: Amsterdam, The Netherlands; Oxford, UK; New York, NY, USA; Tokyo, Japan, 1989; Volume 28, pp. 233–279.
5. Fontana, M.G. *Corrosion Engineering*, 3rd ed.; McGraw-Hill. Inc.: Singapore, 1987.

6. Sandoval-Jabalera, R.; Arias-del Campo, E.; Chacón-Nava, J.G.; Malo-Tamayo, J.M.; Mora-Mendoza, J.L.; Martínez-Villafañe, A. Corrosion Behavior of Engineering Alloys in Synthetic Wastewater. *J. Mater. Eng. Perform.* **2006**, *15*, 53–58. [CrossRef]
7. Rothwell, N.; Tullmin, M. *The Corrosion Monitoring Handbook*, 1st ed.; Coxmoor Publishing Company: Oxfordshire, UK, 2000.
8. Al-Subai, S.G. Corrosion Resistance of Austenitic Stainless Steel in Acetic Acid Solution Containing Bromide Ions. Ph.D. Thesis, University of Manchester, Manchester, UK, 2011.
9. Anaee, R.A.M.; Abdulmajeed, M.H. *Advances in Tribology, Chapter 5 Tribocorrosion*; IntechOpen: Rijeka, Croatia, 2016.
10. Laycock, N.J.; Newman, R.C. Temperature Dependence of Pitting Potentials for Austenitic Stainless Steels above Their Critical Pitting Temperature. *Corrosion Sci.* **1998**, *40*, 887–902. [CrossRef]
11. Park, J.O.; Matsch, S.; Böhni, H. Effects of Temperature and Chloride Concentration on Pit Initiation and Early Pit Growth of Stainless Steel. *J. Electrochem. Soc.* **2002**, *149*, 34–39. [CrossRef]
12. Manning, P.E.; Duquette, D.J. The Effect of Temperature (25–289 °C) on Pit Initiation in Single Phase and Duplex 304l Stainless Steels in 100 ppm Cl^- Solution. *Corrosion Sci.* **1980**, *20*, 597–609. [CrossRef]
13. Carranza, R.M.; Alvarez, M.G. The Effect of Temperature on the Passive Film Properties and Pitting Behavior of a Fe-Cr-Ni Alloy. *Corros. Sci.* **1996**, *38*, 909–925. [CrossRef]
14. Haaz, H.-Y.; Kwon, H.-S. Effects of pH Levels on the Surface Charge and Pitting Corrosion Resistance of Fe. *J. Electrochem. Soc.* **2012**, *159*, C416–C421.
15. Shibata, T.; Zhu, Y.-C. The Effect of Flow Velocity on the Pitting Potential of Anodized Titanium. *Corrosion Sci.* **1995**, *37*, 343–346. [CrossRef]
16. Berradja, A. *Corrosion Inhibitor, Chapter Electrochemical Techniques for Corrosion and Tribocorrosion Monitoring: Fundamentals of Electrolytic Corrosion*; IntechOpen: Rijeka, Croatia, 2019. [CrossRef]
17. Pourbaix, M. *Atlas of Electrochemical Equilibria in Aqueous Solution*; National Association of Corrosion Engineers: Houston, TX, USA, 1974.
18. Zorn, C.; Kaminski, N. Temperature–humidity–bias testing on insulated-gate bipolartransistor modules—Failure modes and acceleration due to high voltage, The Institution of Engineering and Technology. *IET Power Electron.* **2015**, 1–7. [CrossRef]
19. Schweitzer, P.A. *Fundamentals of corrosion—Mechanisms, Causes and Preventative Methods*; Taylor and Francis Group, LLC: Boca Raton, FL, USA, 2010; ISBN 978-1-4200-6770-5.
20. Robert, B. *Corrosion Rate Calculation (from Mass Loss), in Corrosion Tests and Standards: Application and Interpretation*; ASTM International: West Conshohocken, PA, USA, 2005; Volume 23.
21. Hamdy, A.S.; El-Shenawy, E.; El-Bitar, T. Electrochemical Impedance Spectroscopy Study of the Corrosion Behavior of Some Niobium Bearing Stainless Steels in 3.5% NaCl. *Int. J. Electrochem. Sci.* **2006**, *1*, 171–180.
22. Burns, R.M. Electrochemical Techniques in Corrosion Study. *J. Appl. Phys.* **1937**, *8*, 398. [CrossRef]

23. Craig, B.D.; Lane, R.A.; Rose, D.H. *Corrosion Prevention and Control: A Program Management Guide for Selecting Materials*; Advanced Materials, Manufacturing, and Testing Information Analysis Center: Rome, Italy; New York, NY, USA, September 2006.
24. Sharma, S.; Chaudhary, R.S. Inhibitive action of methyl red towards corrosion of mild steel in acids. *Bull. Electrochem.* **2000**, *16*, 267.
25. Bentiss, F.; Traisnel, M.; Lagrenee, M. The substituted 1,3,4-oxadiazoles: A new class of corrosion inhibitors of mild steel in acidic media. *Corrosion Sci.* **2000**, *42*, 127. [CrossRef]
26. Klement, W.; Willens, R.H.; Duwez, P.O.L. Non-crystalline structure in solidified gold-silicon alloys. *Nature* **1960**, *187*, 869–870. [CrossRef]
27. Wang, S. *Metallic Glasses—Formation and Properties, Chapter 4 Corrosion Resistance and Electrocatalytic Properties of Metallic Glasses*; IntechOpen: London, UK, 2016; pp. 63–96. [CrossRef]
28. Scully, J.R.; Gebert, A.; Payer, J.H. Corrosion and related mechanical properties of bulk metallic glasses. *J. Mate. Res.* **2007**, *22*, 302–313. [CrossRef]
29. Huang, R.; Horton, D.J.; Bocher, F.; Scully, J.R. Localized corrosion resistance of Fe-Cr-Mo-W-B-C bulk metallic glasses containing Mn+Si or Y in neutral and acidified chloride solutions. *Corrosion* **2010**, *66*, 035003. [CrossRef]

2. Metallic Glasses and Rapid Solidification

Metallic glasses (MGs) are relatively novel engineering alloys which have attracted widespread interest in terms of fundamental studies as well as a range of applications stemming from thermoplastic processing ability and excellent properties, including high strength, high hardness, corrosion resistance, and soft magnetism [1,2]. They are very unique in terms of their manufacturing routes, atomic structure, processing, properties, and applications.

2.1. Metallic Glass Synthesis

Metallic glasses show amorphous structure with disordered atomic arrangement [1] unlike the periodic distribution of atoms in crystalline metals and alloys. The most common methods for synthesizing metallic glasses include injection molding, suction casting, melt spinning of glassy ribbons, and laser deposition for metallic glass coatings [2,3]. Their synthesis and processing require avoiding crystallization by preventing nucleation and growth of crystalline atomic clusters [4]. At a sufficiently high quench rate of the liquid, crystal nucleation is impeded and an amorphous structure is obtained [5]. Critical cooling rate (R_c) is the slowest cooling rate to avoid crystallization and is one criterion for determining the glass-forming ability (GFA) of a metallic glass former. Alloys with a lower value of R_c have better GFA than those with higher R_c [6]. The early metallic glass compositions required extremely rapid cooling rates (> 10^5 K·s^{-1}) for obtaining amorphous samples. The cooling rate achieved by melt spinning technique is on the order of 10^4–10^6 K·s^{-1}, and metallic glass ribbons may be produced using this method. The melt spinning process involves: (1) induction melting of the alloy in a crucible, (2) forcing molten alloy out of a nozzle located under the crucible onto a rotating wheel, typically made of copper, which is cooled internally, and finally, (3) rapid solidification of the alloy to form metallic glass ribbons (Figure 2.1). Recently developed glass-forming compositions require much lower cooling rates, as low as 1 K·s^{-1}. Some empirical rules for stabilizing the undercooled liquid for successful bypass of crystallization include [4]: (1) alloying of three or more elements; (2) 12% atomic size mismatch between the main constituent elements resulting in "frustration" of a periodic atomic structure; and (3) significantly negative liquid heat of mixing as compared with the corresponding crystalline phase. Generally, a multi-component system with these features is close to the eutectic point in the phase diagram. This allows their fabrication in three-dimensional shapes with thicker cross-sections (up to 10 cm or more) and are known as bulk metallic glasses (BMGs) [5,6].

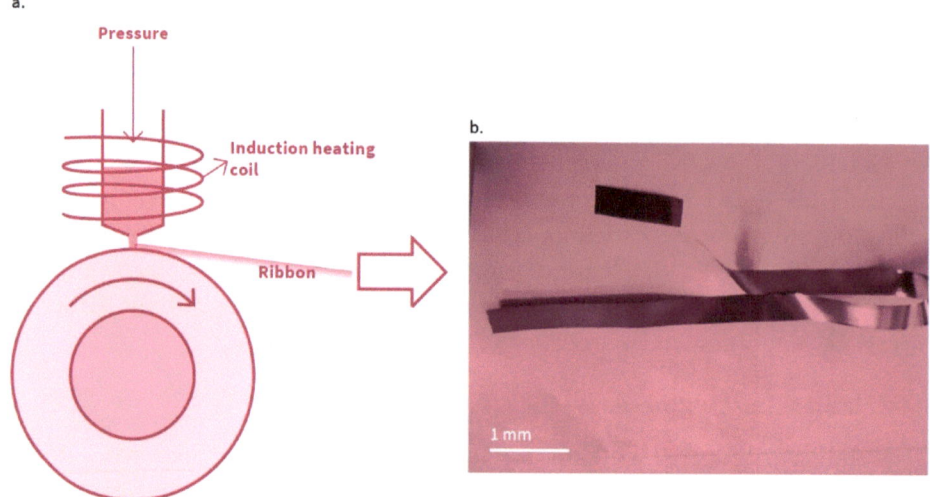

Figure 2.1. (a) Schematic of melt spinning technique to prepare a metallic glass ribbon and (b) a metallic glass ribbon prepared by melt spinning.

2.2. Processing and Microstructure Characterization

Metallic glasses show fully amorphous structure free from microstructural features such as grains, grain boundaries, and secondary phases [7]. X-ray diffraction (XRD), high-resolution transmission electron microscopy (HRTEM), and differential scanning calorimetry (DSC) are frequently used together to characterize amorphous alloys. Figure 2.2 shows the structural characterization and thermal stability of an as-cast $Zr_{57}Nb_5Cu_{15.4}Ni_{12.6}Al_{10}$ alloy as a typical example [8]. X-ray intensity as a function of diffraction angle (2θ) for $Zr_{57}Nb_5Cu_{15.4}Ni_{12.6}Al_{10}$ alloy is shown in Figure 2.2a. A broad diffraction peak is seen, a typical characteristic of the amorphous structure. Figure 2.2b shows random distribution of atoms in the high-resolution transmission electron microscopy (HRTEM) image and diffuse diffraction ring in the selected area diffraction (SAD) pattern for $Zr_{57}Nb_5Cu_{15.4}Ni_{12.6}Al_{10}$ alloy. DSC is typically used to evaluate the characteristic temperatures of an amorphous alloy, including crystallization temperature (T_x) characterized by an exothermic event and glass transition (T_g) by an endothermic event. A typical DSC curve for $Zr_{57}Nb_5Cu_{15.4}Ni_{12.6}Al_{10}$ BMG at a heating rate of 0.33 K·s^{-1} is shown in Figure 2.2c, where T_g and T_x are indicated on the plot.

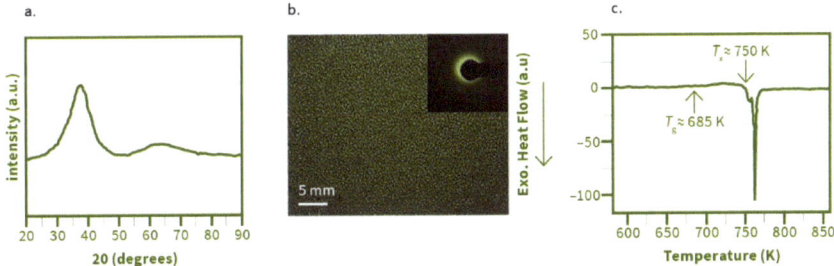

Figure 2.2. Microstructure characterization of $Zr_{57}Nb_5Cu_{15.4}Ni_{12.6}Al_{10}$ bulk metallic glass including: **(a)** XRD pattern with broad diffraction peak, **(b)** HRTEM image with the corresponding selected area diffraction (SAD) pattern shown as an inset and **(c)** DSC profile at constant heating rate of 0.33 K·s^{-1}, showing the glass transition (T_g) and crystallization temperature (T_x) (redrawn with data from reference [8]).

A time-temperature-transformation (TTT) diagram is shown schematically in Figure 2.3 with the critical cooling rate indicated by R_c [9]. Metallic glasses may be thermoplastically processed in the supercooled liquid region bound between T_g and T_x. Metallic glasses soften into a viscous liquid by going above the glass transition temperature and may be net-shaped in a suitable mold with the application of suitable pressure [10–12]. Typical thermoplastic net-shaping of a metallic glass is shown by route (1) in Figure 2.3. Due to this ability, BMGs may be formed into complex geometries and shapes on a wide range of length scales (macro, micro, and nano) not possible using conventional machining methods [10]. Many different shapes and structures have been manufactured utilizing the thermoplastic processing of metallic glasses including nanowires, honeycomb structures, and micro-gears [11–13].

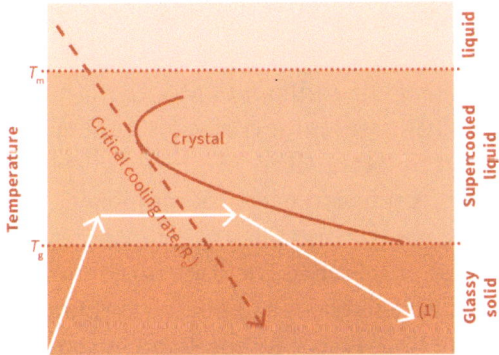

Figure 2.3. Schematic of a time-temperature-transformation (TTT) diagram for a metallic glass-forming alloy between the melting temperature (T_m) and glass transition temperature (T_g). The solid blue arrows show the thermoplastic forming route for a metallic glass above the glass transition temperature (T_g).

2.3. Applications of Metallic Glasses

BMGs have been used in a wide range of structural and functional applications. Due to their superior strength, high elasticity, and good tribological and corrosion characteristics, they are attractive as bio-implants or surgical tools [14,15]. The excellent soft magnetic characteristics of BMGs make them attractive as core materials in high-frequency transformers and magnetic shielding [16,17]. In addition, they are suitable for electronic devices such as surveillance systems, magnetic wires, sensors [18], and energy storage [19,20]. Some other applications of bulk glassy alloys include: (1) airplane slat truck cover [21], (2) casing in cellular phones or electromagnetic instruments [22], (3) high-end industrial components [23], (4) gear parts due to good wear resistance [24], (5) sporting equipment due to large elasticity, (6) mirrors due to surface smoothness, and (7) micro/nano-electromechanical systems (MEMS) because of attractive small-scale properties [7,21,25].

2.4. Corrosion Mechanisms in Metallic Glasses

The superior corrosion resistance of metallic glasses is attributed to their uniform structure free from grain boundaries, dislocations, stacking faults, and intermetallic compounds, which often act as initiation sites for localized corrosion [7,26–28]. The corrosion resistance of some Co-, Ni-, and Fe-based bulk metallic glasses is reported to be two to four orders of magnitude better than the commonly used corrosion-resistant stainless steel (SUS 316L) in various corrosive environments [21]. The chemically homogeneous single-phase nature of amorphous alloys helps in the rapid formation of a highly protective passive film without preferential nucleation sites for corrosion attack [29–31]. In addition to atomic-level homogeneity, the composition of BMGs has an important role in contributing to their corrosion resistance. The addition of certain elements may result in the enhancement of corrosion resistance by the incorporation of protective elements into the passive film or via retarding the dissolution rate of the underlying metallic substrate [28]. Several studies suggest that the addition of Mo, Cr, and tungsten (W) to Fe-P-C metallic glasses drastically increases the corrosion resistance [32–34]. Furthermore, Cr-containing alloys and Zr-based alloys exhibit excellent corrosion resistance in saline environments when compared to crystalline metallic alloys [35,36].

Most of the current reviews on metallic glasses are focused on alloy development, mechanical behavior, and processing. There are no comprehensive reports on the corrosion behavior of amorphous alloys and underlying electrochemical mechanisms. The corrosion behavior of BMGs has a complex and interconnected dependence on chemistry, structure, and surface condition in a range of environments including acidic, basic, and simulated body fluids. In addition, due to the biocompatibility and non-allergic properties of some BMGs, they are attractive for medical components such as prosthetic implants, stents, and surgical tools. This necessitates the evaluation of biocompatibility and biocorrosion response of these alloys. In addition to primary corrosion mechanisms in monolithic BMGs, galvanic corrosion may be seen in BMG composites due to preferential dissolution of the matrix or the crystalline dendrites depending on the chemical partitioning [26]. Since good bulk glass forming compositions typically consist of elements with very different nobility and large difference in standard electrode potentials (e.g., Au-Si, Zr-Cu, or Nb-Ni), they are prone to preferential corrosion if there is elemental partitioning [37,38]. Diffusion in metallic glasses has been reported to be similar to crystalline alloys in some respects, including Arrhenius temperature dependence of diffusion coefficient (D), atomic size effect of diffusing species, and annealing effect of quenched states [39]. On the other hand, diffusion in metallic glasses may be significantly different from their crystalline counterparts based on quantitative measurements due to the presence of free volume and other thermodynamic considerations. Random distribution of atoms in metallic glasses promotes diffusion by cooperative motion of atoms rather than single atom jumps that are commonly observed in crystalline solids. Thus, metallic glasses are less prone to corrosion attacks initiated due to elemental segregation in comparison to their crystalline counterparts. Furthermore, these differences impact the diffusion of corrosive species (including hydrogen) in metallic glasses and affect the observed corrosion kinetics in amorphous alloys [38,39]. Depending on the chemistry, structural state, and processing conditions, the amount of free volume in metallic glasses may be significantly different. Therefore, corrosion mechanisms are likely to be different and unique for each amorphous system and depend on a range of factors including processing, environment, temperature, and pH.

References

1. Jun, W.K.; Willens, R.H.; Duwez, P. Non-crystalline Structure in Solidified Gold–Silicon Alloys. *Nature* **1960**, *187*, 869–870.
2. Jafary-Zadeh, M.; Kumar, G.P.; Branicio, P.S.; Seifi, M.; Cui, J.J.L.a.F. A Critical Review on Metallic Glasses as Structural Materials for Cardiovascular Stent Applications. *J. Funct. Biomater.* **2018**, *9*, 19. [CrossRef]
3. Huang, J.C.; Chu, J.P.; Jang, J.S.C. Recent progress on metallic glasses in Taiwan. *Intermetallics* **2009**, *17*, 973–987. [CrossRef]
4. Suryanarayana, C.; Inoue, A. Iron-based Bulk Metallic Glasses. *Int. Mater. Rev.* **2013**, *58*, 131–166. [CrossRef]
5. Scully, J.R.; Gebert, A.; Payer, J.H. Corrosion and related mechanical properties of bulk metallic glasses. *J. Mater. Res.* **2007**, *22*, 302–313. [CrossRef]
6. Schroeder, V.; Gilbert, C.J.; Ritchie, R.O. Comparison of the Corrosion Behavior of a Bulk Amorphous metal, $Zr_{41.2}Ti_{13.8}Cu_{12.5}Ni_{10}Be_{22.5}$, with its Crystallized Form. *Scripta Mater.* **1998**, *38*, 1481–1485. [CrossRef]

7. Lee, P.Y.; Cheng, Y.M.; Chen, J.Y.; Hu, C.J. Formation and Corrosion Behavior of Mechanically Alloyed Cu-Zr-Ti Bulk Metallic Glasses. *Metals* **2017**, *7*, 148. [CrossRef]
8. Sadeghilaridjani, M.; Ayyagari, A.; Muskeri, S.; Hasannaeimi, V.; Jiang, J.; Mukherjee, S. Small-Scale Mechanical Behavior of Ion-Irradiated Bulk Metallic Glass. *JOM* **2019**, *72*, 1–7. [CrossRef]
9. Miller, M.; Liaw, P.K. *Bulk Metallic Glasses: An Overview*; Springer: New York, NY, USA, 2008.
10. Ma, J.; Liang, X.; Wu, X.; Liu, Z.; Gong, F. Sub-second thermoplastic forming of bulk metallic glasses by ultrasonic beating. *Sci. Rep.* **2015**, *5*, 17844. [CrossRef]
11. Li, N.; Chen, W.; Liu, L. Thermoplastic Micro-Forming of Bulk Metallic Glasses: A Review. *JOM* **2016**, *68*, 1246–1261. [CrossRef]
12. Magagnosc, D.J.; Ehrbar, R.; Kumar, G.; He, M.R.; Schroers, J.; Gianola, D.S. Tunable tensile ductility in metallic glasses. *Sci. Rep.* **2013**, *3*, 1096. [CrossRef]
13. Kumar, G.; Tang, H.X.; Schroers, J. Nanomoulding with amorphous metals. *Nat. Lett.* **2009**, *457*, 868–872. [CrossRef] [PubMed]
14. Li, H.; Zheng, Y.F. Recent advances in bulk metallic glasses for biomedical applications. *Acta Biomater.* **2016**, *36*, 1–20. [CrossRef] [PubMed]
15. Espallargas, N.; Aune, R.E.; Torres, C.; Papageorgiou, N.; Munoz, A.I. Bulk metallic glasses (BMG) for biomedical applications—A tribocorrosion investigation of Zr55Cu30Ni5Al10 in simulated body fluid. *Wear* **2013**, *301*, 27. [CrossRef]
16. Chakri, N.E.; Bendjemil, B.; Baricco, M. Crystallization kinetics and magnetic properties of Fe40Ni40B20 bulk metallic glass. *Adv. Chem. Eng. Sci.* **2014**, *4*, 36. [CrossRef]
17. Ashby, M.F.; Greer, A. Metallic glasses as structural materials. *Scr. Mater.* **2006**, *54*, 321. [CrossRef]
18. Li, H.X.; Lu, Z.C.; Wang, S.L.; Wu, Y.; Lu, Z.P. Fe-based bulk metallic glasses: Glass formation, fabrication, properties and applications. *Prog. Mater. Sci.* **2019**, *103*, 235–318. [CrossRef]
19. Tiberto, P.; Baricco, M.; Olivetti, E.; Piccin, R. Magnetic properties of bulk metallic glasses. *Adv. Eng. Mater.* **2007**, *9*, 468. [CrossRef]
20. Nowosielski, R.; Babilas, R.; Ochin, P.; Stoklosa, Z. Thermal and magnetic properties of selected Fe-based metallic glasses. *Arch. Mater. Sci. Eng.* **2008**, *30*, 13.
21. Inoue, A.; Wang, X.M.; Zhang, W. Development and Application of Bulk Metallic Glasses. *Rev. Adv. Mater. Sci.* **2008**, *18*, 1–9.
22. Hofmann, D.C.; Roberts, S.N. Microgravity metal processing: From undercooled liquids to bulk metallic glasses. *NPJ Microgr.* **2015**, *1*, 15003. [CrossRef]
23. Schroers, J. Bulk Metallic Glasses. *Phys. Today* **2013**, *66*, 32. [CrossRef]
24. Developing Ceramic-Like Bulk Metallic Glass Gears. NASA's Jet Propulsion Laboratory. Available online: https://www.techbriefs.com/component/content/article/tb/techbriefs/manufacturingprototyping/21647 (accessed on 15 December 2020).
25. Chen, M. Brief overview of bulk metallic glasses. *NPG Asia Mater.* **2011**, *3*, 82–90. [CrossRef]
26. Qiao, J.; Fan, J.; Yang, F.; Shi, X.; Lan, H.Y.a.A. The Corrosion Behavior of Ti-based Metallic Glass Matrix Composites in the H_2SO_4 Solution. *Metals* **2018**, *8*, 52. [CrossRef]

27. Virtanen, S.; Böhni, H. Passivity, breakdown and repassivation of glassy Fe-Cr-P alloys. *Corrosion Sci.* **1990**, *31*, 333–342. [CrossRef]
28. Weller, K. Thermodynamics and kinetics of the oxidation of amorphous Al-Zr alloys. *Max-Planck-Inst. Intell. Syst.* **2015**. [CrossRef]
29. Lee, H.J.; Akiyama, E.; Habazaki, H.; Kawashima, A.; Hashimoto, K.A.a.K. The Corrosion Behavior of Amorphous and Crystalline Ni-10Ta-20P Alloys in 12 M HCl. *Corrosion Sci.* **1996**, *38*, 1269–1279. [CrossRef]
30. Tabeshian, A.; Persson, D.; Arnberg, L.; Aune, R.E. Corrosion behavior of bulk amorphous and crystalline Zr-based alloys in simulated body fluid with and without additions of protein. *Mater. Corros.* **2016**, *67*, 748–755. [CrossRef]
31. Naka, M.; Hashimoto, K.; Masumoto, T. Change in Corrosion Behavior of Amorphous Fe-P-C Alloys by Alloying with Various Metallic Elements. *J. Non-Cryst. Solids* **1979**, *31*, 355–365. [CrossRef]
32. Pang, S.J.; Zhang, T.; Asami, K.; Inoue, A. New Fe–Cr–Mo–(Nb, Ta)–C–B glassy alloys with high glass-forming ability and good corrosion resistance. *Mater. Trans.* **2001**, *42*, 376–379. [CrossRef]
33. Qin, C.; Zhang, W.; Kimura, H.; Asami, K.; Inoue, A. New Cu-Zr-Al-Nb Bulk Glassy Alloys with High Corrosion Resistance. *Mater. Trans.* **2004**, *45*, 1958–1961. [CrossRef]
34. Pang, S.; Zhang, T.; Asami, K.; Inoue, A. Formation of Bulk Glassy Ni–(Co–)Nb–Ti–Zr Alloys with High Corrosion Resistance. *Mater. Trans.* **2002**, *43*, 1771–1773. [CrossRef]
35. Kawashima, A.; Wada, T.; Ohmura, K.; Xie, G.; Inoue, A. A Ni- and Cu-free Zr-based bulk metallic glass with excellent resistance to stress corrosion cracking in simulated body fluids. *Mater. Sci. Eng. A* **2012**, *542*, 140–146. [CrossRef]
36. Morrison, M.; Buchanan, R.; Liaw, P.; Green, B.; Wang, G.; Liu, C.; Horton, J. Corrosion–fatigue studies of the Zr-based Vitreloy 105 bulk metallic glass. *Mater. Sci. Eng. A* **2007**, *467*, 198–206. [CrossRef]
37. Stanojevic, S.; Gallino, I.; Aboulfadl, H.; Sahin, M.; Mücklich, F.; Busch, R. Oxidation of glassy Ni–Nb–Sn alloys and its influence on the thermodynamics and kinetics of crystallization. *Acta Mater.* **2016**, *102*, 176–186. [CrossRef]
38. Eisenbart, M.; Klotz, U.; Busch, R.; Gallino, I. A colourimetric and microstructural study of the tarnishing of gold-based bulk metallic glasses. *Corros. Sci.* **2014**, *85*, 258–269. [CrossRef]
39. Faupel, F.; Frank, W.; Macht, M.P.; Mehrer, H.; Naundorf, V.; Rätzke, K.; Schober, H.R.; Sharma, S.K.; Teichler, H. Diffusion in metallic glasses and supercooled melts. *Rev. Mod. Phys.* **2003**, *75*, 237–280. [CrossRef]

3. Zirconium (Zr)-based Bulk Metallic Glasses and Their Composites

3.1. Zr-based Bulk Metallic Glasses

Zr-based bulk metallic glasses (BMGs) generally show excellent glass-forming ability (GFA) with low critical cooling rates. They have been the subject of numerous studies due to their promising properties and processing ability [1–9]. Some of the attractive attributes of Zr-based BMGs include a low Young's modulus, high fatigue limit, high elasticity and strength, and good wear resistance. Biocompatibility and good formability make these alloys attractive for load-bearing implants such as artificial joints and dental prostheses [6] in complex geometries. In addition, Zr-based BMGs show similar or better cell adhesion and proliferation compared to commercial biomedical alloys [10].

3.2. Corrosion Behavior of Zr-Based Metallic Glasses

The corrosion behavior of Zr-based BMGs has been widely studied in a range of neutral, acidic, and alkaline electrolytes. Specifically, in terms of biocorrosion behavior, several investigations were performed for Zr-Al-Ni-Cu and Zr-Ti-Ni-Cu-Be alloy systems known for their excellent glass-forming ability and good mechanical properties [10]. Biocompatibility and electrochemical behavior were evaluated for $Zr_{65}Cu_{17.5}Ni_{10}Al_{7.5}$ BMG in phosphate-buffered solution (PBS) under different conditions of pH, surface finish, and Cl^- concentration [11]. This amorphous system exhibited good corrosion resistance. Cell viability studies for another Zr-based BMG (Zr-10Al-5Ti-17.9Cu-14.6Ni) showed good biocompatibility similar to that of Ti [12]. $Zr_{41.2}Ti_{13.8}Ni_{10}Cu_{12.5}Be_{22.5}$ BMG (commonly referred to as Vitreloy 1) demonstrated better corrosion resistance than 316L stainless steel in PBS that was comparable to Co- and Ti-alloys [13]. However, for a biomaterial to be of practical use in the human body, toxic elements such as Ni and beryllium (Be) need to be avoided to prevent allergic reactions, despite the advantages of these elements in terms of corrosion resistance enhancement [14]. To that end, a series of Ni-free and Be-free Zr-based BMGs were developed with good GFA, biocompatibility, and cell viability albeit with lower corrosion resistance [15–18]. Elements with a high level of toxicity to cellular metabolism such as Ni, Be, and Cu may limit the use of alloy systems with these constituents as biomaterials [19,20]. Ni causes allergic reactions [21], has an anti-proliferative effect on cells [22], and is carcinogenic to the human body [23]. The high biocorrosion resistance of the alloys may prevent metallic ions from being released into the human body, thus making them safer for use [24–26]. A number of Ni-free Zr-based BMGs have been developed recently to eliminate possible allergic response including Zr-Al-Cu [27], Zr-Al-Co-(Cu) [11, 12,28], Zr-Al-Co-Nb [29,30], Zr-Al-Co-Ag [31], Zr-Al-Cu-Fe [32], Zr-Al-Cu-Ag [33], Zr-Cu-Pd-Al-Nb [34], and Zr-Al-Cu-Fe-(Ti/Nb) [35].

Several methods have been utilized for improvement of the corrosion behavior of Zr-based BMGs including changing of the surface composition or microstructure through micro-arc oxidation, ion-implantation, or other surface treatments [36,37]; lowering of free volume [38]; introduction of a second phase into the amorphous matrix by annealing or in situ synthesis [39]; and the addition of elements with strong passivation characteristics [40,41]. Some of the recently developed Zr-based BMGs exhibit excellent GFA, a wide super-cooled liquid region, and excellent biocorrosion resistance in simulated body fluid such as Zr-Nb-Cu-Pd-Al, Zr-Ti-Cu-Fe-Al, Zr-Nb-Cu-Fe-Al, and Zr-Co-Al-Ag [33].

3.3. Effect of Alloying Elements

Alloying elements typically used for improving passivation resistance in metallic alloys are classified as dissolution moderators and passivity promoters [42]. Dissolution moderators or blockers such as Mo, Nb, Ta, and W are known to form strong metal–metal bonds and reduce the dissolution of elements from the surface in corrosive electrolytes. On the other hand, passivity promoters such as Ti, Cr, and Pt form weak metal–metal bonds but typically promote a high degree of oxygen adsorption, resulting in strong passive films on the surface. More active elements might be dissolved and leached from the surface in a corrosive environment with the enrichment of passivity promoters on the surface.

3.3.1. Effect of Copper (Cu) Addition

Cu-containing Zr-based BMGs have been studied extensively for their corrosion behavior and biocompatibility [33]. One of the most studied Zr-based amorphous alloy systems is Zr-Cu-Al ($Zr_{50}Cu_{43}Al_7$) which does not show good pitting resistance in SBF [34]. This was attributed to the formation of nanocrystals on the surface, which accelerates dissolution by galvanic corrosion. High Cu content adversely affects the corrosion resistance and pitting behavior of Zr-based BMGs [43–46]. For the alloy system Zr–Cu–Al–(Ni–Nb, Ni–Ti, Ag) (Cu = 15.4–36 at. %), the re-passivation potential (E_R) and pitting potential (E_P) decreased almost linearly with increase in Cu content, as shown in Figure 3.1, although the corrosion potential (E_{corr}) was roughly unchanged [43]. Pit nucleation and growth in these BMGs begin with the selective dissolution of Zr, Al, and other valve-metals, which results in Cu enrichment inside the pits. Subsequently, Cu reacts with Cl^- ions to form CuCl, which turns into Cu_2O after hydrolysis [47]. The presence of local Cu-rich regions may lead to galvanic cells and accelerate dissolution of the glassy phase. This may explain the observed low re-passivation ability in these BMGs [45,46,48–50]. The detrimental effect of increasing Cu content on the corrosion behavior of Zr-based metallic glasses has also been reported for other electrolytes including PBS [51], H_2SO_4, HCl, and NaCl [52]. Binary Zr-Cu metallic glasses with 40–60 at. % Cu showed lower pitting potential compared to pure Zr [53,54]. The presence of regions enriched with Cu beneath the passive layer and the formation of chloride compounds inside the pits result in weakening of the surface protective film [55].

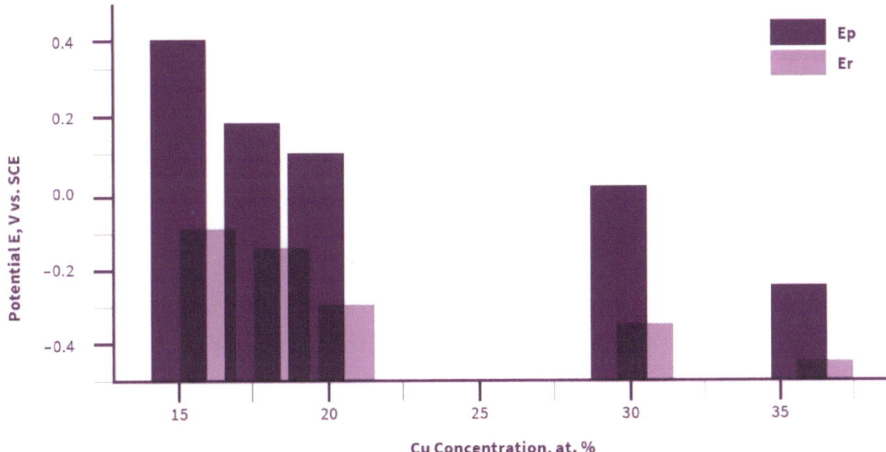

Figure 3.1. Pitting (E_p) and re-passivation potentials (E_r) of five Zr-Cu-based bulk metallic glasses (BMGs; Zr–Cu–Al–(Ni–Nb, Ni–Ti, Ag) (Cu = 15.4–36 at. %)) including $Zr_{52.5}Cu_{17.9}Ni_{14.6}Al_{10}Ti_5$ (Vit 105) and $Zr_{58.5}Cu_{15.6}Ni_{12.8}Al_{10.3}Nb_{2.8}$ (Vit 106) in 0.01 M Na_2SO_4 + 0.1 M NaCl solution (redrawn data from reference [43]).

Correlations between glass-forming ability, width of the super-cooled liquid region (ΔT_x), and corrosion behavior has been systematically investigated for Zr-based BMGs with varying Cu content [56]. The alloy with highest Cu content showed the lowest values of ΔT_x and E_{pit} and highest i_{corr} in Cl^--containing solutions. This was attributed to the increasing chemical inhomogeneity with higher fraction of Cu. The increase in Zr/Cu ratio resulted in the enhancement of corrosion resistance and surface wettability of Zr-Cu-Al-Nb-Pd BMGs [57]. For the higher Zr/Cu ratio in the Zr-Cu-Al-Fe system, better biocorrosion resistance and cell adhesion compared to 316L stainless steel was reported, indicating high stability in the in vivo environment [58]. XPS results showed the formation of a stable ZrO_2-rich surface layer and increase in Cu content (more than or equal to ~ 20 at. %) improved glass-forming ability [59]. Pit initiation and growth in BMGs with high Cu content have been attributed to chemical and physical heterogeneities on the surface that lead to the selective dissolution of active metals such as Zr, Al, and Ni and local enrichment of Cu in the pits [46,47,60,61]. Many pits were reported on the surface of $Zr_{50}Cu_{40}Al_{10}$ BMG after immersion in NaCl, while almost no pits were found for $Zr_{70}Cu_6Al_8Ni_{16}$ with lower copper content in the same solution after long immersion time [60].

3.3.2. Effects of Niobium (Nb) and Cobalt (Co) Addition

Co and Nb have been reported to increase the corrosion resistance of Zr-based BMGs attributed to the formation of highly protective passive films on the surface [29, 30]. Increasing Co content in the $(Zr_{48}Cu_{46}Al_8)_{100-x}Co_x$ alloy system from 0 to 4 at. % significantly reduced the corrosion rate and increased the pitting potential [62]. Substitution of Co with Nb in $Zr_{55}Al_{20}Co_{25-x}Nb_x$ (x = 0, 2.5, and 5 at. %) and $Zr_{56}Al_{16}Co_{28-x}Nb_x$ (x= 1, 3, and 5 at. %) alloy systems led to widening of the passive region and an increase in pitting potential in PBS and Hank's solution [63,64]. The effect of alloying elements such as Titanium (Ti), Chromium (Cr), Niobium (Nb), or Tantalum (Ta) on the corrosion behavior of amorphous Zr–Al–Ni–Cu alloys in HCl and NaCl environments showed that Nb had the largest effect on corrosion resistance followed by Ta, Ti, and Cr [65]. Specifically, the effect of Nb addition on corrosion performance was investigated for the Zr-Nb-Cu-Ni-Al alloy system in simulated body fluid [66] and for the Zr-Nb-Cu-Ag-Al alloy system in NaCl, HCl, and H_2SO_4 [67]. Nb shifted the corrosion potential to more positive values and increased the pitting potential. With the addition of Nb, release of harmful elements such as Ni and Cu reduced dramatically due to the formation of protective passive film in artificial blood plasma (ABP) [4]. The improvement in corrosion resistance with the addition of a small amount of Nb has been reported in different environments for Zr-based BMG systems including $Zr_{59}Cu_{20}Al_{10}Ni_8Nb_3$ [28], $(Cu_{60}Zr_{30}Ti_{10})_{95}Nb_5$ [29], $Zr_{55}Al_{20-x}Co_{25}Nb_x$ (x= 0 to 5 at. %) [30], $(Zr_{56}Al_{16}Co_{28})_{100-x}Nb_x$ (x= 0, 2, and 4) [68], and $Zr_{46-x}Nb_xCu_{37.6}Ag_{8.4}Al_8$ (x= 0, 0.5, 1, 2, and 4) [67]. A small addition of Nb increased the pitting potential and passive film stability in Ni-free Zr-based bulk metallic glasses [69]. XPS depth analysis demonstrated that Nb addition promoted the oxidation of Zr and resulted in thicker oxide films [70]. This is similar to Hafnium (Hf) addition that promoted the formation of passive films and enhanced the corrosion resistance of Zr-Cu-Ni-Al BMGs in different electrolytes [71]. Zr-Al-Co BMGs have been reported to have excellent corrosion resistance due to the high fraction of Zr and Co, both the elements promoting passivity in different media [72].

3.3.3. Effect of Silver (Ag) Addition

The antibacterial nature of Ag has long been known and therefore, incorporated in numerous biomaterials including orthopedic implants [73], dental casting alloys [74], bioactive glasses [75], and other applications [76]. Ag was introduced in some Zr-based amorphous alloys to improve glass-forming ability [76]. The addition of Ag in the Zr-Al-Co alloy system resulted in increased corrosion resistance in PBS [76]. Silver containing Zr-Cu-Al-Ag amorphous alloy system shows excellent corrosion resistance in NaCl solution and Hank's solution [77], which was attributed to the formation of a homogenous Al_2O_3-enriched passive film resulting from the presence of Ag. Cell culture studies indicated that the presence of Ag improved biocompatibility of Ag-containing amorphous alloys. The addition of Ag to Zr-Co-Al amorphous ternary system increased the pitting potential of the BMG in NaCl and enhanced E_{corr} in H_2SO_4 [78] due to the formation of Zr- and Al-rich passive film [6,79].

3.3.4. Effect of Rare-Earth (RE) Elements Addition

The addition of rare-earth elements such as yttrium (Y) has been shown to improve the corrosion resistance of Zr-based BMGs. Electrochemical tests for $Zr_{55}Al_{10}Cu_{30}Ni_5$ and $(Zr_{55}Al_{10}Cu_{30}Ni_5)_{99}Y_1$ alloys in phosphate-buffered saline (PBS) electrolyte showed that Y enhanced the biocorrosion resistance of Zr-Al-Cu-Ni BMGs by accelerating the formation of surface passive film [80]. Simultaneous friction and wear have been shown to accelerate $Zr_{55}Al_{10}Cu_{30}Ni_5$ corrosion due to removal of the surface passive film in SBF [81]. Increase of Y content in the $(Zr_{58}Nb_3Cu_{16}Ni_{13}Al_{10})_{100-x}Y_x$ alloy system increased the pitting corrosion susceptibility of Zr-based BMGs in NaCl solution due to precipitation of copper-yttrium clusters that acted as heterogeneities [82]. Therefore, the addition of yttrium affected the corrosion behavior of Zr-based BMGs in different ways.

3.4. Combined Effects of Mechanical Loading and Corrosion

Zr-Ti-Cu-Ni-Be amorphous alloys have been shown to be susceptible to stress corrosion cracking and stress corrosion fatigue in NaCl solution due to the formation of non-protective oxide at the crack tip [83]. Corrosion fatigue life and the endurance limit for $Zr_{52.5}Cu_{17.9}Ni_{14.6}Al_{10}Ti_5$ (Vit 105) in NaCl solution were shown to be similar to many Al alloys [84]. The degradation mechanism was found to be anodic dissolution rather than hydrogen embrittlement, showing a significant decrease in open-circuit potential (OCP) during the fatigue test when the crack growth accelerated [84]. Corrosive environment did not affect the fatigue life of $(Zr_{0.55}Al_{0.10}Ni_{0.05}Cu_{0.30})_{99}Y_1$ BMG in the higher stress range while it was significantly more pronounced at lower stresses and decreased fatigue strength of the alloy by about 40% [8].

3.5. Effects of Structure and Crystallinity

The effect of thermally induced structural changes (e.g., crystallization or relaxation) on electrochemical properties is of significant interest in engineering applications of bulk amorphous alloys. Partial or complete crystallization may deteriorate or improve the corrosion resistance of a metallic glass depending on the specific composition. There are several reports on the corrosion behavior of Zr-based metallic glasses after crystallization or relaxation. $Zr_{56}Co_{16}Al_{28}$ [85] metallic glass showed increased corrosion rate in Ringer's simulated body fluid after heat treatment (i.e., crystallization) with a less stable surface passive film as compared to its amorphous counterpart. A similar trend was observed for other Zr-based amorphous alloys such as $Zr_{55}Cu_{30}Ni_5Al_{10}$ [86], $Zr_{53}Cu_{30}Ni_9Al_8$ [87], $Zr_{52.5}Cu_{17.9}Ni_{14.6}Al_{10}Ti_5$ [50], and $Zr_{65}Al_{7.5}Cu_{17.5}Ni_{10}$ [88]. Crystallized $Zr_{41.25}Ti_{13.75}Ni_{10}Cu_{12.5}Be_{22.5}$ alloys did not show good passivation characteristics during polarization tests in NaCl solution because of chemically inhomogeneous microstructure and crystalline defects [89]. The corrosion behavior of certain Zr-based metallic glasses improved after partial or complete crystallization due to easier formation of the surface passive film. As an example, $Zr_{60}Cu_{20}Al_{10}Fe_5Ti_5$ and $Zr_{62.3}Cu_{22.5}Fe_{4.9}Al_{6.8}Ag_{3.5}$ alloys demonstrated lower corrosion current density and better resistance to pitting after crystallization in SBF and simulated seawater, respectively [90,91]. Similar to crystallization, relaxation by thermal annealing has been reported to change the corrosion behavior of Zr-based BMGs. $Zr_{52.5}Cu_{17.9}Ni_{14.6}Al_{10}Ti_5$ bulk metallic glass was obtained in three different structural states including as-cast, mechanically deformed, and relaxed with different amounts of free volume. For the thermally relaxed sample with lower free volume content, the corrosion resistance and pit propagation resistance increased [92]. On the other hand, the deformed sample with relatively higher amount of free volume exhibited lower corrosion resistance. Cold-rolling of $Zr_{62}Cu_{12.5}Ni_{10}Al_{7.5}Ag_8$ and friction stir processing (FSP) of Zr-Ti-Cu-Ni-Be BMGs have been reported to decrease the corrosion resistance of these alloys due to increase in free volume [93,94]. Similar behavior was reported for $Zr_{60}Cu_{20}Al_{10}Fe_5Ti_5$ [95], $Zr_{41.2}Ti_{13.8}Cu_{12.5}Ni_{10}Be_{22.5}$ [96], and $Zr_{57}Cu_{15.4}Ni_{12.6}Al_{10}Nb_5$ [96] BMGs, where corrosion resistance improved after relaxation. A smaller amount of free volume improves corrosion resistance due to the formation of dense amorphous structure with low residual stresses and more stable and protective passive film [97].

3.6. Zr-Based Bulk Metallic Glasses Composites

Metallic glass matrix composites (MGMCs) have been developed over the last two decades as a new strategy for enhancing the plasticity and toughness of bulk metallic glasses [98–100]. They consist of a crystalline dendritic phase in the amorphous matrix, leading to significantly greater plasticity compared to monolithic glasses by preventing catastrophic propagation of shear bands [98–100]. While there are numerous reports on the microstructure and mechanical behavior of MGMCs, there are very limited number of studies on their corrosion behavior. The mechanism of passive film formation in MGMCs is very similar to the fully amorphous systems [49,71,101–103]. Rapid pit propagation and a sharp increase in current density in H_2SO_4 electrolyte for potentials higher than E_{pit} have been reported for $Zr_{66.64}Nb_{6.4}Cu_{10}Ni_{8.7}Al_{8.0}$ alloy consisting of body centered cubic (BCC) dendrites in a glassy matrix and also for the $Zr_{57}Ti_8Nb_{2.5}Cu_{13.9}Ni_{11.1}Al_{7.5}$ alloy with quasi-crystals and glassy intergranular phase [71]. This behavior is similar to that reported for some Zr-based BMGs [104]. Both the composites mentioned above showed chloride-induced selective dissolution of the glassy matrix phase, while the crystalline phase remained un-attacked after polarization [49].

Scanning electrochemical microscopy (SECM) was used to evaluate the localized electrochemical activity of Zr-based MGMCs [105]. The interface between the amorphous matrix and the dendritic crystalline phase acted as sites for preferential corrosion [106]. In chloride-containing solutions (HCl and NaCl), $Zr_{58.5}Ti_{14.3}Nb_{5.2}Cu_{6.1}Ni_{4.9}Be_{11.0}$ MGMC showed preferential dissolution of the amorphous matrix, while uniform corrosion was observed for the same MGMC in NaOH [106]. In Cl^--containing solutions, the favorable absorption of chloride ions at the interface between the matrix and the crystalline phase resulted in selective dissolution of the amorphous matrix. In contrast, the dissolution of beryllium oxide was the primary mechanism for protective film breakdown in other solutions such as H_2SO_4 and NaOH [106]. The increase in defect density in the vicinity of the phase boundaries has also been reported to promote pitting corrosion in MGMCs [71].

3.7. Effect of Test Environment

Environmental factors such as chemistry and pH of the solution, temperature, and presence of oxygen influence the corrosion behavior of metallic materials. A number of Zr-based BMGs including $Zr_{65}Cu_{17.5}Ni_{10}Al_{7.5}$, $Zr_{60}Nb_5Cu_{17.5}Ni_{10}Al_{7.5}$, and $Zr_{60}Nb_5Cu_{17.5}Ni_5Pd_5Al_{7.5}$ were evaluated in different types of artificial body fluids including artificial saliva solution (ASS), phosphate-buffered solution (PBS), and artificial blood plasma (ABP) solution, and exhibited excellent corrosion resistance in these solutions, making them promising as biomaterials [26].

Guan et al. [63] investigated the role of different electrolytes on the corrosion resistance of Zr-Al-Co-Nb BMGs. This alloy system passivated in NaCl solution with a low passive current density. However, the alloys showed higher resistance to pitting corrosion in Hank's solution compared to NaCl solution. The passivation region was even wider for the alloys in PBS. This behavior was attributed to the complex composition of Hank's solution and PBS containing large-sized PO_4^{3-} anions, which were absorbed on the surface and inhibited pitting. These anions compete with Cl^- and reduce the aggressive nature of Cl^-, resulting in a delay in the active–passive process in PBS and Hank's solution [63]. The presence of amino acids and proteins in the solution has also been reported to act as diffusion barrier to other species such as phosphate and chloride ions and improve the corrosion resistance of Zr-BMGs [107]. $Zr_{41.2}Ti_{13.8}Ni_{10}Cu_{12.5}Be_{22.5}$ BMGs exhibited the lowest corrosion current density (i_{corr})

in H_2SO_4 followed by NaCl and HNO_3, and the highest i_{corr} was obtained in the HCl environment [90,108]. In terms of acid concentration, $Zr_{59}Ti_3Cu_{20}Al_{10}Ni_8$ metallic glass displayed a lower passivation tendency at higher nitric acid concentration (1, 6, and 11.5 M HNO_3) [109]. A similar trend was observed with increasing chloride concentration for the $Zr_{65}Al_{7.5}Ni_{10}Cu_{17.5}$ amorphous alloy [20]. On the contrary, a higher concentration of dissolved oxygen (4% O_2) in PBS solution led to higher corrosion resistance for the $Zr_{65}Al_{7.5}Ni_{10}Cu_{17.5}$ alloy in a neutral pH [11].

A change in electrolyte temperature has also been reported to influence the corrosion behavior of Zr-based BMGs. With the increase in temperature, the passive film formed on the $Zr_{55}Cu_{30}Al_{10}Ni_5$ alloy became thicker, less stable, and more porous compared to the films formed at room temperature. Increasing temperature generally led to a decrease in pitting potential and higher pitting susceptibility [110]. Finally, a change in temperature has been reported to change the electrolyte pH with an adverse effect on corrosion behavior [86]. The corrosion current density (i_{corr}), corrosion potential (E_{corr}), and pitting potential (E_{pit}) of Zr-based BMGs as a function of environment and test temperature are summarized in Table 3.1 below.

Table 3.1. Corrosion parameters for Zr-based BMGs in different environments.

Zr-Based BMG	i_{corr} ($\mu A/cm^2$)	E_{corr} (mV vs. SCE)	E_{pit} (mV vs. SCE)	Environment	T (°C)	Ref.
Simulated Body Fluid						
$Zr_{62.3}Cu_{22.5}Fe_{4.9}Al_{6.8}Ag_{3.5}$	0.055	−290	−22	PBS (phosphate-buffered saline solution)	37	[2]
$Zr_{62.3}Cu_{22.5}Fe_{4.9}Al_{6.8}Ag_{3.5}$ (crystalline)	0.077	−305	−145	PBS	37	[2]
$Zr_{55}Al_{10}Ni_5Cu_{30}$	0.54	−231	118	PBS	37	[80]
$(Zr_{55}Al_{10}Ni_5Cu_{30})_{99}Y_1$	0.043	−302	185	PBS	37	[80]
$Zr_{60}Cu_{22.5}Fe_{7.5}Al_{10}$	0.201	−370	50	PBS	37	[111]
$Zr_{60}Cu_{20}Fe_{10}Al_{10}$	0.264	−390	70	PBS	37	[111]
$Zr_{60}Ti_{20}Cu_{10}Fe_5Al_{10}$	0.234	−420	280	PBS	37	[111]
$Zr_{60}Ti_{2.5}Fe_{12.5}Al_{10}Cu_{10}Ag$	0.123	−340	282	PBS	-	[10]
$Zr_{60}Ti_{2.5}Fe_{10}Al_{10}Cu_{10}Ag_5$	0.111	−261	432	PBS	-	[10]
$Zr_{65}Ti_{2.5}Fe_{7.5}Al_{10}Cu_{10}Ag_5$	0.086	−184	484	PBS	-	[10]
$Zr_{60.14}Cu_{22.31}Fe_{4.85}Al_{9.7}Ag_3$	0.014	−416	917	SBF	37	[33]
$Zr_{50}Cu_{43}Al_7$	0.23	−410	125	Artificial Saliva (AS)	37	[35]
$Zr_{50}Cu_{43}Al_7$	0.21	−550	−162	SBF	37	[35]
$Zr_{56}Al_{16}Co_{28}$	-	−635	459	HBSS (Hank's balanced saline solution)	37	[72]
$Zr_{56}Al_{16}Co_{28}$	-	−580	368	PBS	37	[72]
$Zr_{57}Nb_5Cu_{15.4}Ni_{12.6}Al_{10}$	0.014	−382	291	SBF	37	[112]
$Zr_{57}Nb_5Cu_{15.4}Ni_{12.6}Al_{10}$	0.038	−263	786	AS	37	[112]
$Zr_{62.3}Cu_{22.5}Fe_{4.9}Al_{6.8}Ag_{3.5}$	0.055	−290	−22	PBS	-	[2]
$Zr_{53}Al_{15}(Co_{1-x}Ag_x)_{31}$ (x = 0, 0.1, 0.25)	-	−530 to −400	30 to 225	PBS	-	[6]
$Zr_{50}Cu_{35}Al_7Nb_5Pb_3$	-	<−300	<350	PBS	-	[57]
$Zr_{55}Cu_{30}Al_7Nb_5Pb_3$	-	<−300	<600	PBS	-	[57]
$Zr_{62.5}Cu_{22.5}Fe_5Al_{10}$	0.03	−532	787	AS	-	[34]
$Zr_{62.5}Cu_{22.5}Fe_5Al_{10}$	0.02	−537	531	SBF	-	[34]
$Zr_{53}Cu_{30}Ni_9Al_8$	0.06	−214	35	SBF	-	[87]
$Zr_{46}(Cu_{4.5/5.5}Ag_{1/5.5})_{46}Al_8$	4.24	−555	-	Hank's	37	[77]

Table 3.1. Cont.

Zr-Based BMG	i_{corr} (μA/cm^2)	E_{corr} (mV vs. SCE)	E_{pit} (mV vs. SCE)	Environment	T (°C)	Ref.
$Zr_{51.9}Cu_{23.3}Ni_{10.5}Al_{4.3}$	4.61	−575	-	Hank's	37	[77]
$Zr_{51}Ti_5Ni_{10}Cu_{25}Al_9$	9.03	−646	-	Hank's	37	[77]
$Ti_{40}Zr_{25}Ni_{12}Cu_3Be_{20}$	10.09	−625	-	Hank's	37	[77]
$Zr_{52.5}Cu_{17.9}Ni_{14.6}Al_{10.0}Ti_{5.0}$ (Vit 105)	-	−405	69	PBS	-	[113]
NaCl Solution						
$Zr_{56}Al_{16}Co_{28}$	-	−642	269	0.9% NaCl	37	[72]
$Zr_{46}(Cu_{4.5/5.5}Ag_{1/5.5})_{46}Al_8$	0.123	−388	-	0.9% NaCl	37	[77]
$Zr_{51.9}Cu_{23.3}Ni_{10.5}Al_{4.3}$	0.135	−411	-	0.9% NaCl	37	[77]
$Zr_{51}Ti_5Ni_{10}Cu_{25}Al_9$	0.137	−425	-	0.9% NaCl	37	[77]
$Ti_{40}Zr_{25}Ni_{12}Cu_3Be_{20}$	0.361	−404	-	0.9% NaCl	37	[77]
$Zr_{41.2}Ti_{13.8}Ni_{10}Cu_{12.5}Be_{22.5}$	0.672	−496	-	3.5% NaCl	25	[108]
$Zr_{35}Ti_{30}Be_{35}$	-	−445	84.5	0.6 M NaCl	25	[114]
$Zr_{35}Ti_{30}Be_{29}Co_6$	-	−424	257	0.6 M NaCl	25	[114]
$Zr_{52.5}Cu_{17.9}Ni_{14.6}Al_{10}Ti_5$	-	−264	324	0.6 M NaCl	25	[114]
$Zr_{41.25}Ti_{13.75}Ni_{10}Cu_{12.5}Be_{22.5}$	-	−194	<−120	3.5% NaCl	-	[89]
$Zr_{52.5}Cu_{17.9}Ni_{14.6}Al_{10.0}Ti_{5.0}$						
Deformed	6.49	−291.4	64			
As cast	4.74	−289	54.3	0.6 M NaCl		[92]
Relaxed	1.98	−308.7	−43.2			
$Zr_{55}Cu_{30}Ni_5Al_{10}$						
As cast	1.709	−107.2				
Annealed at 360 °C	11.401	−276.5		3.5% NaCl	-	[115]
Annealed at 400 °C	6.761	−113.2				
Annealed at 480 °C	5.411	−204.4				
$Zr_{58.3}Al_{14.6}Ni_{8.3}Cu_{18.8}$	2.5	−153.92	−20.27	3% NaCl		[56]
$Zr_{58}Al_{16}Ni_{11}Cu_{15}$	1.96	−167.95	−22	3% NaCl	-	[56]
$Zr_{57.5}Al_{17.5}Ni_{13.8}Cu_{11.3}$	2.03	158.08	−12.3	3% NaCl		[56]
$Zr_{56}Al_{16}Ni_{28}$	0.096	−298	−33	3% NaCl	25	[116]
$Zr_{56}Al_{16}(Ni_{0.7}Ag_{0.3})_{28}$	0.087	−258	168	3% NaCl	25	[116]
$Zr_{55}Al_{10}Ni_5Ni_{30}$	0.177	−388	−175	3% NaCl	25	[116]
$Zr_{60}Al_{10}Ni_{10}Ni_{20}$	0.118	−355	−75	3% NaCl	25	[116]
$Zr_{46}Al_8Cu_{38}Ag_8$	0.126	−330	−223	3% NaCl	25	[116]
$Zr_{50}Cu_{40}Al_{10}$	-	−495	−433	0.5 M NaCl	30	[60]
$Zr_{70}Cu_6Al_8Ni_{16}$	-	−376	−70	0.5 M NaCl	30	[60]
Other Solutions						
$Zr_{59}Ti_3Cu_{20}Al_{10}Ni_8$	55	680 (vs. Ag/AgCl)	1400	1 N HNO$_3$	25	[55]
$Zr_{59}Ti_3Cu_{20}Al_{10}Ni_8$	140	630	1300	6 N HNO$_3$	25	[55]
$Zr_{59}Ti_3Cu_{20}Al_{10}Ni_8$	300	570	1100	11.5 N HNO$_3$	25	[55]
$Zr_{41.2}Ti_{13.8}Ni_{10}Cu_{12.5}Be_{22.5}$	0.672	−496		3.5% NaCl	25	[108]
$Zr_{41.2}Ti_{13.8}Ni_{10}Cu_{12.5}Be_{22.5}$	0.899	−428		1 N HNO$_3$	25	[108]
$Zr_{41.2}Ti_{13.8}Ni_{10}Cu_{12.5}Be_{22.5}$	0.539	−491		1 N H$_2$SO$_4$	25	[108]
$Zr_{41.2}Ti_{13.8}Ni_{10}Cu_{12.5}Be_{22.5}$	1.373	−322		1 N HCl	25	[108]

References

1. Wang, W.; Dong, C.; Shek, C. Bulk metallic glasses. *Mater. Sci. Eng. R* **2004**, *44*, 45–89. [CrossRef]
2. Guo, S.; Liu, Z.; Chan, K.C.; Chen, W.; Zhang, H.J.; Wang, J.F.; Yu, P. A plastic Ni-free Zr-based bulk metallic glass with high specific strength and good corrosion properties in simulated body fluid. *Mater. Lett.* **2012**, *84*, 81–84. [CrossRef]
3. Liu, L.; Qiu, C.; Sun, M.; Chen, Q.; Chan, K.; Pang, G.K. Improvements in the plasticity and biocompatibility of Zr–Cu–Ni–Al bulk metallic glass by the microalloying of Nb. *Mater. Sci. Eng. A* **2007**, *449–451*, 193–197. [CrossRef]
4. Wang, Y.; Liu, Y.; Liu, L. Fatigue Behaviors of a Ni-free ZrCuFeAlAg Bulk Metallic Glass in Simulated Body Fluid. *J. Mater. Sci. Technol.* **2014**, *30*, 622–626. [CrossRef]

5. Hua, N.; Zheng, Z.; Fang, H.; Ye, X.; Lin, C.; Li, G.; Wang, W. Dry and lubricated tribological behavior of a Ni- and Cu-free Zr-based bulk metallic glass. *J. Non-Cryst. Solids* **2015**, *426*, 63–71. [CrossRef]
6. Hua, N.; Huang, L.; Wang, J.; Cao, Y.; He, W.; Pang, S.; Zhang, T. Corrosion behavior and in vitro biocompatibility of Zr–Al–Co–Ag bulk metallic glasses: An experimental case study. *J. Non-Cryst. Solids* **2012**, *358*, 1599–1604. [CrossRef]
7. Qiu, C.; Chen, Q.; Liu, L.; Chan, K.; Zhou, J.; Chenc, P.; Zhang, A.S. A novel Ni-free Zr-based bulk metallic glass with enhanced plasticity and good biocompatibility. *Scr. Mater.* **2006**, *55*, 605–608. [CrossRef]
8. Huang, L.; Wang, G.; Qiao, D.; Liaw, P.K.; Pang, S.; Wang, J.; Zhang, T. Corrosion-fatigue study of a Zr-based bulk-metallic glass in a physiologically relevant environment. *J. Alloy. Comp.* **2010**, *504S*, S159–S162. [CrossRef]
9. Greer, A.L.; Ma, E. Bulk metallic glasses: at the cutting edge of metals research. *MRS Bull.* **2007**, *32*, 611–619. [CrossRef]
10. Hua, N.; Huang, L.; Chen, W.; He, W.; Zhang, T. Biocompatible Ni-free Zr-based bulk metallic glasses with high-Zr-content: Compositional optimization for potential biomedical applications. *Mater. Sci. Eng. C* **2014**, *44*, 400–410. [CrossRef]
11. Hiromoto, S.; Tsai, A.; Sumita, M.; Hanawa, T. Effect of surface finishing and dissolved oxygen on the polarization behavior of $Zr_{65}Al_{7.5}Ni_{10}Cu_{17.5}$ amorphous alloy in phosphate buffered solution. *Corrs. Sci.* **2000**, *42*, 2167–2185. [CrossRef]
12. Horton, J.; Parsell, D. Biomedical potential of a zirconium-based bulk metallic glass. *Mater. Res. Soc. Symp. Proc.* **2003**, *754*, CC1.5.1–CC1.5.16. [CrossRef]
13. Morrison, M.L.; Buchanan, R.A.; Peker, A.; Peter, W.H.; Horton, J.A.; Liaw, P.K. Cyclic-anodic-polarization studies of a $Zr_{41.2}Ti_{13.8}Ni_{10}Cu_{12.5}Be_{22.5}$ bulk metallic glass. *Intermetallics* **2004**, *12*, 1177–1181. [CrossRef]
14. Li, Y.; Zhang, W.; Dong, C.; Qiang, J.; Fukuhara, M.; Makino, A.; Inoue, A. Effects of Ni addition on the glass-forming ability, mechanical properties and corrosion resistance of Zr–Cu–Al bulk metallic glasses. *Mater. Sci. Eng. A* **2011**, *528*, 8551–8556. [CrossRef]
15. Zberg, B.; Uggowitzer, P.; Loffler, J. MgZnCa glasses without clinically observable hydrogen evolution for biodegradable implants. *Nat. Mater.* **2009**, *8*, 887–891. [CrossRef]
16. Jin, K.; Löffler, J.F. Bulk metallic glass formation in Zr–Cu–Fe–Al alloys. *Appl. Phys. Lett.* **2005**, *86*, 241909. [CrossRef]
17. Buzzi, S.; Jin, K.; Uggowitzer, P.; Tosatti, S.; Gerber, I.; Löffler, J. Cytotoxicity of Zr based bulk metallic glasses. *Intermetallics* **2006**, *14*, 29–34. [CrossRef]
18. Wessels, V.; Mené, G.L.; Fischerauer, S.; Kraus, T.; Weinberg, A.M.; Uggowitzer, P. In vivo performance and structural relaxation of biodegradable bone implants made from MgZnCa bulk metallic glasses. *Adv. Eng. Mater.* **2012**, *14*, 357–364. [CrossRef]
19. Wang, J.; Choi, B.; Nieh, T.; Liu, C. Nano-scratch Behavior of a Bulk Zr–10Al–5Ti–17.9Cu–14.6Ni Amorphous Alloy. *J. Mater. Res.* **2000**, *15*, 913–922. [CrossRef]

20. Hiromoto, S.; Tsai, A.; Sumita, M.; Hanawa, T. Effect of chloride ion on the anodic polarization behavior of the $Zr_{65}Al_{7.5}Ni_{10}Cu_{17.5}$ amorphous alloy in phosphate buffered solution. *Corros Sci.* **2000**, *42*, 1651–1660. [CrossRef]
21. Kalimo, K.; Mattila, L.; Kautiainen, H. Nickel allergy and orthodontic treatment. *J. Eur. Acad. Dermatol. Venereol.* **2004**, *18*, 543–545. [CrossRef] [PubMed]
22. Wataha, J.; Lockwood, P.; Schedle, A. Effect of silver, copper, mercury, and nickel ions on cellular proliferation during extended, low-dose exposures. *J. Biomed. Mater. Res.* **2000**, *52*, 360–364. [CrossRef]
23. Sunderman, F. Carcinogenic effects of metals. *Fed. Proc.* **1978**, *37*, 40–46. [PubMed]
24. Schroers, J.; Kumar, G.; Hodges, T.; Chan, S.; Kyriakides, T. Bulk metallic glasses for biomedical applications. *JOM* **2009**, *61*, 21–29. [CrossRef]
25. Demetriou, M.; Wiest, A.; Hofmann, D.; Johnson, W.; Han, B.; Wolfson, N.; Wang, G.; Liaw, P. Amorphous metals for hard-tissue prosthesis. *JOM* **2010**, *62*, 83–91. [CrossRef]
26. Liu, L.; Qiu, C.; Chen, Q.; Zhang, S. Corrosion behavior of Zr-based bulk metallic glasses in different artificial body fluids. *J. Alloy. Comp.* **2006**, *425*, 268–273. [CrossRef]
27. Hiromoto, S.; Tsai, A.; Sumita, M.; Hanawa, T. Effect of pH on the polarization behavior of $Zr_{65}Al_{7.5}Ni_{10}Cu_{17.5}$ amorphous alloy in a phosphate-buffered solution. *Corros. Sci.* **2000**, *42*, 2193–2200. [CrossRef]
28. Raju, V.; Kühn, U.; Wolff, U.; Schneider, F.; Eckert, J.; Reiche, R.; Gebert, A. Corrosion behaviour of Zr-based bulk glass-forming alloys containing Nb or Ti. *Mater. Lett.* **2002**, *57*, 173–177. [CrossRef]
29. Qin, C.; Asami, K.; Zhang, T.; Zhang, W.; Inoue, A. Corrosion Behavior of Cu-Zr-Ti-Nb Bulk Glassy Alloys. *Mater. Trans.* **2003**, *44*, 749–753. [CrossRef]
30. Pang, S.; Zhang, T.; Asami, K.; Inoue, A. Formation, corrosion behavior, and mechanical properties of bulk glassy Zr–Al–Co–Nb alloys. *J. Mater. Res.* **2003**, *18*, 1652–1658. [CrossRef]
31. Inoue, T.; Zhang, J.; Saida, M.; Matsushita, M.; Sakurai, T. Formation of Icosahedral Quasicrystalline Phase in Zr–Al–Ni–Cu–M (M=Ag, Pd, Au or Pt) Systems. *Mater. Trans. JIM* **1999**, *40*, 1181–1184. [CrossRef]
32. Liu, L.; Chan, K.; Pang, G. The microprocesses of the quasicrystalline transformation in $Zr_{65}Ni_{10}Cu_{7.5}Al_{7.5}Ag_{10}$ bulk metallic glass. *Appl. Phys. Lett.* **2004**, *85*, 2788–2790. [CrossRef]
33. Liu, Y.; Wang, Y.-M.; Pang, H.-F.; Zhao, Q.; Liu, L. A Ni-free ZrCuFeAlAg bulk metallic glass with potential for biomedical applications. *Acta Biomater.* **2013**, *9*, 7043–7053. [CrossRef]
34. Huang, H.-H.; Huang, H.-M.; Lin, M.-C.; Zhang, W.; Sun, Y.-S.; Kai, W. Enhancing the bio-corrosion resistance of Ni-free ZrCuFeAl bulk metallic glass through nitrogen plasma immersion ion implantation. *J. Alloy. Comp.* **2014**, *615*, S660–S665. [CrossRef]
35. Huang, H.H.; Sun, Y.S.; Wu, C.P.; Liu, C.F.; Liaw, P.K.; Kai, W. Corrosion resistance and biocompatibility of Ni-free Zr-based bulk metallic glass for biomedical applications. *J. Int.* **2012**, *30*, 139–143.

36. Liu, L.; Liu, Z.; Chan, K.; Luo, H.; Cai, Q.; Zhang, S. Surface modification and biocompatibility of Ni-free Zr-based bulk metallic glass. *Scr. Mater.* **2008**, *58*, 231–234. [CrossRef]
37. Jiang, Q.; Qin, C.; Amiya, K.; Nagata, S.; Inoue, A.; Zheng, R.; Cheng, G.; Nie, X.; Jiang, J. Enhancement of corrosion resistance in bulk metallic glass by ion implantation. *Intermetallics* **2008**, *16*, 225–229. [CrossRef]
38. Schroeder, V.; Ritchie, R. Stress-corrosion fatigue-crack growth in a Zr-based bulk amorphous metal. *Acta Mater.* **2006**, *54*, 1785–1794. [CrossRef]
39. Kawashima, A.; Kurishita, H.; Kimura, H.; Inoue, A. Effect of chloride ion concentration on the fatigue crack growth rate of a $Zr_{55}Al_{10}Ni_5Cu_{30}$ bulk metallic glass. *Mater. Trans.* **2007**, *48*, 1969–1972. [CrossRef]
40. Kawashima, A.; Yokoyama, Y.; Inoue, A. Zr-based bulk glassy alloy with improved resistance to stress corrosion cracking in sodium chloride solutions. *Corros. Sci.* **2010**, *52*, 2950–2957. [CrossRef]
41. Yokoyama, Y.; Fujita, K.; Yavari, A.; Inoue, A. Malleable hypoeutectic Zr–Ni–Cu–Al bulk glassy alloys with tensile plastic elongation at room temperature. *Philos. Mag. Lett.* **2009**, *89*, 322–334. [CrossRef]
42. Marcus, P. On some fundamental factors in the effect of alloying elements on passivation of alloys. *Corros. Sci.* **1994**, *36*, 2155–2158. [CrossRef]
43. Gostin, P.F.; Eigel, D.; Grell, D.; Eckert, J.; Kerscher, E.; Gebert, A. Comparing the pitting corrosion behavior of prominent Zr-based bulk metallic glasses. *J. Mater. Res.* **2015**, *30*, 233–241. [CrossRef]
44. Li, Y.; Zhang, W.; Qin, F.; Makino, A. Mechanical properties and corrosion resistance of a new $Zr_{56}Ni_{20}Al_{15}Nb_4Cu_5$ bulk metallic glass with a diameter up to 25 mm. *J. Alloy. Comp.* **2014**, *615*, S71–S74. [CrossRef]
45. Mudali, U.K.; Baunack, S.; Eckert, J.; Schultz, L.; Gebert, A. Pitting corrosion of bulk glass-forming zirconium-based alloys. *J. Alloys Compd.* **2004**, *377*, 290–297. [CrossRef]
46. Green, B.; Meyer, H.; Benson, R.; Yokoyama, Y.; Liaw, P.; Liu, C. A study of the corrosion behaviour of $Zr_{50}Cu_{(40-X)}Al_{10}Pd_X$ bulk metallic glasses with scanning Auger microanalysis. *Corros. Sci.* **2008**, *50*, 1825–1832. [CrossRef]
47. Gebert, P.G.; Schultz, L. Effect of surface finishing of a Zr-based bulk metallic glass on its corrosion behaviour. *Corros. Sci.* **2010**, *52*, 1711–1720. [CrossRef]
48. Homazava, N.; Shkabko, A.; Logvinovich, D.; Krähenbühl, U.; Ulrich, A. Element specific in situ corrosion behaviour of Zr–Cu–Ni–Al–Nb bulk metallic glass in acidic media studied using a novel microcapillary flow injection inductively coupled plasma mass spectrometry technique. *Intermetallics* **2008**, *16*, 1066–1072. [CrossRef]
49. Gebert, A.; Kuehn, U.; Baunack, S.; Mattern, N.; Schultz, L. Pitting corrosion of zirconium-based bulk glass–matrix composites. *Mater. Sci. Eng. A* **2006**, *415*, 242–249. [CrossRef]
50. Peter, W.; Buchanan, R.; Liu, C.; Liaw, P.; Morrosion, M.; Horton, J.C., Jr.; Wright, J. Localized corrosion behaviour of a zirconium-based bulk metallic glass relative to its crystalline state. *Intermetallics* **2002**, *10*, 1157–1162. [CrossRef]

51. Long, M.; Rack, H. Titanium alloys in total joint replacement—A materials science perspective. *Biomaterials* **1998**, *19*, 1621–1639. [CrossRef]
52. Li, Y.; Zhang, W.; Dong, C.; Qiang, J.; Xie, G.; Fujita, K.; Inoue, A. Glass-forming ability and corrosion resistance of Zr-based Zr–Ni–Al bulk metallic glasses. *J. Alloys Comp.* **2012**, *536S*, S117–S121. [CrossRef]
53. Varano, R.; Bobyn, J.; Medley, J.; Yue, S. Effect of microstructure on the dry sliding friction behavior of CoCrMo alloys used in metal-on-metal hip implants. *J. Biomed. Mater. Res. Part B Appl. Biomater.* **2006**, *76*, 281–286. [CrossRef] [PubMed]
54. Lewandowski, J.; Greer, A. Temperature rise at shear bands in metallic glasses. *Nat. Mater.* **2006**, *5*, 15–18. [CrossRef]
55. Padhy, N.; Ningshen, S.; Mudali, U.K. Electrochemical and surface investigation of zirconium based metallic glass $Zr_{59}Ti_3Cu_{20}Al_{10}Ni_8$ alloy in nitric acid and sodium chloride media. *J. Alloys Comp.* **2010**, *503*, 50–56. [CrossRef]
56. Tauseefa, A.; Tariqa, N.; Akhter, J.; Hasan, B.; Mehmood, M. Corrosion behavior of Zr–Cu–Ni–Al bulk metallic glasses in chloride medium. *J. Alloys Comp.* **2010**, *489*, 596–599. [CrossRef]
57. Huang, L.; Yokoyama, Y.; Wu, W.; Liaw, P.; Pang, S.; Inoue, A.; Zhang, T.; Wei, H. Ni-free Zr–Cu–Al–Nb–Pd bulk metallic glasses with different Zr/Cu ratios for biomedical applications. *J. Biomed. Mater. Res.* **2012**, *100B*, 1472–1482. [CrossRef]
58. Huang, L.; Pu, C.; Fisher, R.K.; Mountain, D.J.; Gao, Y.; Liaw, P.K.; Zhang, W.; He, W. A Zr-based bulk metallic glass for future stent applications: Materials properties, finite element modeling, and in vitro human vascular cell response. *Acta Biomater.* **2015**, *25*, 356–368. [CrossRef]
59. Niinomi, M. Recent metallic materials for biomedical applications. *Metall. Mater. Trans. A* **2002**, *33A*, 477–486. [CrossRef]
60. Kawashima, K.O.; Yokoyama, Y.; Inoue, A. The corrosion behaviour of Zr-based bulk metallic glasses in 0.5 M NaCl solution. *Corros. Sci.* **2011**, *53*, 2778–2784. [CrossRef]
61. Green, B.; Steward, R.; Kim, I.; Choi, C.; Liaw, P.; Kihm, K.; Yokoyama, Y. In-situ observation of pitting corrosion of the $Zr_{50}Cu_{40}Al_{10}$ bulk metallic glass. *Intermetallics* **2009**, *17*, 568–571. [CrossRef]
62. Zhou, W.; Weng, W.P.; Hou, J. Glass-forming Ability and Corrosion Resistance of Zr—Cu—Al—Co Bulk Metallic Glass. *J. Mater. Sci. Technol.* **2016**, *32*, 349–354. [CrossRef]
63. Guan, B.; Shi, X.; Dan, Z.; Xie, G.; Niinomi, M.; Qin, F. Corrosion behavior, mechanical properties and cell cytotoxity of Zr-based bulk metallic glasses. *Intermetallics* **2016**, *72*, 69–75. [CrossRef]
64. Lu, X.; Huang, L.; Zhang, S.J.P.a.T. Formation and biocorrosion behavior of Zr-Al-Co-Nb bulk metallic glasses. *Mater. Sci.* **2012**, *57*, 1723–1727. [CrossRef]
65. Köster, U.; Zander, D. Corrosion of amorphous and nanocrystalline Zr-based alloys. *Mater. Sci. Eng. A* **2004**, *375–377*, 53–59.
66. Qiu, C.; Liu, L.; Sun, M.; Zhang, S. The effect of Nb addition on mechanical properties, corrosion behavior, and metal-ion release of ZrAlCuNi bulk metallic glasses in artificial body fluid. *J. Biomed. Mater. Res.* **2005**, *75*, 950–956. [CrossRef]

67. Nie, X.; Xu, X.; Jiang, Q.; Chen, L.; Xua, Y.; Fang, Y.; Xie, G. Effect of microalloying of Nb on corrosion resistance and thermal stability of ZrCu-based bulk metallic glasses. *J. Non-Cryst. Solids* **2009**, *355*, 203–207. [CrossRef]
68. Le, W.K.; Yuan, Z.Z.; Zhang, X.Y. Experimental and theoretical study on the corrosion resistance of Zr–Co–Al–Nb metallic glasses. *J. Theor. Appl. Phys.* **2017**, *11*, 37–43. [CrossRef]
69. Liu, L.; Qiu, C.L.; Chen, Q.; Zhang, K.C.C.a.S.M. Deformation behavior, corrosion resistance, and cytotoxicity of Ni-free Zr-based bulk metallic glasses. *J. Biomed. Mater. Res. Part A* **2008**, *86*, 160–169. [CrossRef]
70. Cao, Q.; Peng, S.; Zhao, X.; Wang, X.; Zhang, D.; Jiang, J. Effect of Nb substitution for Cu on glass formation and corrosion behavior of ZreCueAgeAleBe bulk metallic glass. *J. Alloys Comp.* **2016**, *683*, 22–31. [CrossRef]
71. Liu, L.; Qiu, C.; Zou, H.; Chan, K. The effect of the microalloying of Hf on the corrosion behavior of ZrCuNiAl bulk metallic glass. *J. Alloys Comp.* **2005**, *399*, 144–148. [CrossRef]
72. Kawashima, T.; Wada, K.O.; Xie, G.; Inoue, A. A Ni- and Cu-free Zr-based bulk metallic glass with excellent resistance to stress corrosion cracking in simulated body fluids. *Materi. Sci. Eng. A* **2012**, *542*, 140–146. [CrossRef]
73. Wan, Y.; Raman, S.; He, F.; Huang, A.Y. Surface modification of medical metals by ion implantation of silver and copper. *Vacuum* **2007**, *81*, 1114. [CrossRef]
74. Wataha, J. Biocompatibility of dental casting alloys: A review. *J. Prosthet. Dent.* **2000**, *83*, 223. [CrossRef]
75. Chen, W.; Liu, Y.; Courtney, H.; Bettenga, M.; Agrawal, C.; Bumgardner, J.D.; Ong, A.J. In vitro anti-bacterial and biological properties of magnetron co-sputtered silver-containing hydroxyapatite coating. *Biomaterials* **2006**, *27*, 5512. [CrossRef] [PubMed]
76. Hua, N.; Pang, S.; Li, Y.; Wang, J.; Li, R. Ni- and Cu-free Zr–Al–Co–Ag bulk metallic glasses with superior glass-forming ability. *J. Mater. Res.* **2011**, *26*, 539–546. [CrossRef]
77. Sun, Y.; Huang, Y.; Fan, H.; Wang, Y.; Ning, Z.; Liu, F. In vitro and in vivo biocompatibility of an Ag-bearing Zr-based bulk metallic glass for potential medical use. *J. Non-Cryst. Solids* **2015**, *419*, 82–91. [CrossRef]
78. Long, Z.; Shen, B.; Shao, Y.; Chang, C.; Zeng, Y.; Inoue, A. Corrosion Behaviour of [(Fe$_{0.6}$Co$_{0.4}$)$_{0.75}$B$_{0.2}$Si$_{0.05}$]$_{96}$Nb$_4$ Bulk Glassy Alloy in Sulphuric Acid Solutions. *Mater. Trans. JIM* **2006**, *47*, 2566. [CrossRef]
79. Zhang, C.; Li, N.; Pan, J.; Guo, S.; Zhang, M.; Liu, A.L. Enhancement of glasses-forming ability and bio-corrosion resistance of Zr-Co-Al bulk metallic glasses by the addition of Ag. *J. Alloys Compd.* **2010**, *504S*, 163. [CrossRef]
80. Huang, L.; Qiao, D.; Green, B.A.; Liaw, P.K.; Wang, J.; Pang, S.; Zhang, T. Bio-corrosion study on zirconium-based bulk-metallic glasses. *Intermetallics* **2009**, *17*, 195–199. [CrossRef]
81. Espallargas, N.; Aune, R.E.; Torres, C.; Papageorgiou, N.; Muno, A.I.Z. Bulk metallic glasses (BMG) for biomedical applications—A tribocorrosion investigation of Zr$_{55}$Cu$_{30}$Ni$_5$Al$_{10}$ in simulated body fluid. *Wear* **2013**, *301*, 271–279. [CrossRef]

82. Yu, L.; Tang, J.; Qiao, J.; Wang, H.; Wang, Y.; Apreutesei, M.; Duan, M.C.A.M. Effect of Yttrium Addition on Corrosion Resistance of Zr-based Bulk Metallic Glasses in NaCl Solution. *Int. J. Electrochem. Sci.* **2017**, *12*, 6506–6519. [CrossRef]
83. Schroers, J.; Nguyen, T.; O'Keeffe, S.; Desai, A. Thermoplastic forming of bulk metallic glass: applications for MEMS and microstructure fabrication. *J. Mater Sci Eng C* **2007**, *449–451*, 898–902. [CrossRef]
84. Jayaraj, J.; Gebert, A.; Schultz, L. Passivation behaviour of structurally relaxed $Zr_{48}Cu_{36}Ag_8Al_8$ metallic glass. *J. Alloys Compd.* **2009**, *479*, 257–261. [CrossRef]
85. Li, H.F.; Zheng, Y.F.; Xu, F.; Jiang, J.Z. In vitro investigation of novel Ni free Zr-based bulk metallic glasses as potential biomaterials. *J. Mater. Lett.* **2012**, *75*, 74–76. [CrossRef]
86. Monfared, H.V.; Glasses, A.S.F.B.M. Biocorrosion and biocompatibility of Zr–Cu–Fe–Al. *Surf. Interface Anal.* **2013**, *45*, 1714–1720. [CrossRef]
87. Liu, Z.; Huang, L.; Wu, W.; Luo, X.; Shi, M.; Liaw, P.K.; He, W.; Zhang, T. Novel low Cu content and Ni-free Zr-based bulk metallic glasses for biomedical applications. *J. Non-Cryst. Solids* **2013**, *363*, 1–5. [CrossRef]
88. Wiest, A.; Wang, G.; Huang, L.; Roberts, S.; Demetriou, M.D.; Liaw, P.K.; Johnson, W.L. Corrosion and corrosion fatigue of Vitreloy glasses containing low fractions of late transition metals. *Scr. Mater.* **2010**, *62*, 540–543. [CrossRef]
89. Nie, X.; Yang, X.; Chen, L.; Yeap, K.; Zeng, K.; Li, D.; Pan, J.S.; Wang, X.; Cao, Q.; Ding, S.; Jiang, J. The effect of oxidation on the corrosion resistance and mechanical properties of a Zr-based metallic glass. *Corros. Sci.* **2011**, *53*, 3557–3565. [CrossRef]
90. Tatschl, A.; Gilbert, C.J.; Schroeder, V.; Pippan, R.; Ritchie, R.O. Stereophotogrammetric investigation of overload and cyclic fatigue fracture surface morphologies in a Zr–Ti–Ni–Cu–Be bulk metallic glass. *Mater. Res.* **2000**, *15*, 898–903. [CrossRef]
91. Morrison, M.; Buchanan, R.; Liaw, P.; Green, B.; Wang, G.; Liu, C.; Horton, J. Corrosion–fatigue studies of the Zr-based Vitreloy 105 bulk metallic glass. *Mater. Sci. Eng. A* **2007**, *467*, 198–206. [CrossRef]
92. Wang, G.Y.; Liaw, P.; Yokoyama, Y.; Inoue, A.; Liu, C. Fatigue behavior of Zr-based bulk-metallic glasses. *Mater. Sci. Eng. A* **2008**, *494*, 314–323. [CrossRef]
93. Chen, P.Y. Fatigue behavior of high-entropy alloys: A review. *Sci. China Technol. Sci.* **2017**, *61*, 168–178. [CrossRef]
94. Wang, R.; Wang, Y.; Yang, J.; Xiong, J.S.a.L. Influence of heat treatment on the mechanical properties, corrosion behavior, and biocompatibility of $Zr_{56}Al_{16}Co_{28}$ bulk metallic glass. *J. Non-Cryst. Solids* **2015**, *411*, 45–52. [CrossRef]
95. Tabeshian, A. Production, Characterization and Electrochemical Properties of Advanced Bulk Metallic Glasses for Hip Implant Applications. MA thesis, Norwegian University of Science and Technology, Trondheim, Norway, 2011. Available online: http://kth.diva-portal.org/smash/record.jsf?pid=diva2%3A559274&dswid=-7337 (accessed on 20 December 2020).
96. Huang, C.; Huang, J.; Li, J.; Jang, J. Simulated body fluid electrochemical response of Zr-based metallic glasses with different degrees of crystallization. *Mater. Sci. Eng. C* **2013**, *33*, 4183–4187. [CrossRef] [PubMed]

97. Li, W.; Su, H.L.; Yue, J. Effects of crystallization on corrosion resistance and electron work function of $Zr_{65}Al_{7.5}Cu_{17.5}Ni_{10}$ amorphous alloys. *J. Philos. Mag. Lett.* **2013**, *93*, 130–137. [CrossRef]
98. Kou, H.-C.; Li, Y.; Zhang, T.-B.; Li, J.; Li, J.-S. Electrochemical corrosion properties of Zr- and Ti-based bulk metallic glasses. *Transf. Nonferrous Metals Soc. China* **2011**, *21*, 552–557. [CrossRef]
99. Wang, S. Corrosion Resistance and Electrocatalytic Properties of Metallic Glasses. *Metall. Glasses Form. Prop.* **2016**. [CrossRef]
100. Guo, S.F.; Zhang, H.J.; Liu, Z.; Xie, W.C.a.S.F. Corrosion resistances of amorphous and crystalline Zr-based alloys in simulated seawater. *Electrochem. Commun.* **2012**, *24*, 39–42. [CrossRef]
101. Jiang, W.H.; Jiang, F.; Green, B.A.; Liu, F.X.; Liaw, a.P.K. Electrochemical corrosion behavior of a Zr-based bulk-metallic glass. *Appl. Phys. Lett.* **2007**, *91*, 1–3. [CrossRef]
102. Zhou, W.; Sheng, J.X.H.a.M.Q. Microstructural Change and Corrosion Behavior During Rolling of Zr-based Bulk Metallic Glass. *Int. J. Electrochem. Sci.* **2016**, *11*, 7163–7172. [CrossRef]
103. Ayyagari, V.A.; Mukherjee, H.S.A.a.S. Corrosion behavior of ZrTiCuNiBe bulk metallic glass subjected to friction stir processing. *J. Non-Cryst. Solids* **2015**, *425*, 124–129.
104. González, S.; Pellicer, E.; Suriñach, S.; Sort, M.D.B.a.J. Mechanical and corrosion behaviour of as-cast and annealed $Zr_{60}Cu_{20}Al_{10}Fe_5Ti_5$ bulk metallic glass. *Intermetallics* **2012**, *28*, 149–155. [CrossRef]
105. Ayyagari, A.; Wu, H.F.; Arora, H.; Mukherjee, S. Amorphous Metallic Alloys: Pathways for Enhanced Wear and Corrosion Resistance. *JOM* **2017**, *69*, 2150–2155.
106. Raicheff, R.; Zaprianova, V.; Gattef, E. Effect of structural relaxation on electrochemical corrosion behaviour of amorphous alloys. *J. Mater. Sci. Lett.* **1997**, *16*, 1701–1704. [CrossRef]
107. Choi-Yim, H.; Johnson, W. Bulk metallic glass matrix composites. *Appl. Phys. Lett.* **1997**, *71*, 3808. [CrossRef]
108. Choi-Yim, H.; Conner, R.; Szecs, F.; Johnson, W. Quasistatic and dynamic deformation of tungsten reinforced $Zr_{57}Nb_5Al_{10}Cu_{15.4}Ni_{12.6}$ bulk metallic glass matrix composites. *Scripta Mater.* **2001**, *45*, 1039–1045. [CrossRef]
109. Hays, C.; Kim, C.; Johnson, W. Microstructure Controlled Shear Band Pattern Formation and Enhanced Plasticity of Bulk Metallic Glasses Containing in situ Formed Ductile Phase Dendrite Dispersions. *Phys. Rev. Lett.* **2000**, *84*, 2901. [CrossRef]
110. Zhang, T.; Lin, X.-M.; Hui, J.-Y.; Liu, Y.; Wang, Y.-S.; Qiao, J.-W. Tribological Properties of a Dendrite-reinforced Ti-based Metallic Glass Matrix Composite under Different Conditions. *J. Iron Steel Res. Int.* **2016**, *23*, 57–63. [CrossRef]
111. Rahaman, M.; Zhang, L. On the estimation of interface temperature during contact sliding of bulk metallic glass. *Wear* **2014**, *320*, 77–86. [CrossRef]
112. Jaeger, J. Moving Sources of Heat and the Temperature of Sliding Contacts. *Proc. R. Soc.* **1942**, *76*, 203–224.

113. Gebert, A.; Mudali, U.K.; Eckert, J.; Schultz, L. Electrochemical reactivity of zirconium-based bulk metallic glasses. *Mater. Res. Soc. Symp. Proc.* **2004**, *806*, 369.
114. Ayyagari, A.; Hasannaeimi, V.; Arora, H.; Mukherjee, S. Electrochemical and Friction Characteristics of Metallic Glass Composites at the Microstructural Length-scales. *Sci. Rep.* **2018**, *8*, 1–10.
115. Tiana, H.; Qiao, J.; Yang, H.; Wang, Y.; Liaw, P.; Lan, A. The corrosion behavior of in-situ Zr-based metallic glass matrix composites in different corrosive media. *Appl. Surface Sci.* **2016**, *363*, 37–43. [CrossRef]
116. Hiromoto, S.; Tsai, A.P.; Masae, S.; Hanawa, T. Polarization Behavior of Bulk Zr-Base Amorphous Alloy Immersed in Cell Culture Medium. *Mat. Trans.* **2002**, *43*, 3112–3117. [CrossRef]

4. High-Density Metallic Glasses

Among all the metallic glass systems studied in terms of their corrosion behavior, the Fe-, Ni-, and Co-based bulk metallic glass (BMG) forming systems are the most widely reported. The corrosion behavior of these three alloy systems depends on their composition, structure, and test environment. The Fe-, Ni-, and Co-based alloy systems are merged together with Cu- and Cr-based BMGs under the topic of high-density metallic glasses.

4.1. Iron (Fe)-Based Metallic Glasses

One of the first reports on the electrochemical behavior of amorphous melt-spun Fe-Cr-P-C ribbons was published in 1974, showing their high corrosion resistance in HCl, even better than that of 304 stainless steel [1,2]. Subsequently, the exceptional corrosion resistance for amorphous Fe-Cr-Mo-C-B and Fe-Ni-Cr-Mo-B alloys specifically in the acidic environment was reported [3,4]. Fe-based BMGs are attractive due to their unique combination of high strength and hardness, high glass transition temperature (T_g), good wear resistance, soft magnetic properties, and good corrosion resistance [5]. The corrosion resistance of Fe-based BMGs was further enhanced with the addition of alloying elements such as Cr and Mo. Similar to stainless steel, Cr was found to be most influential in the improvement of their corrosion resistance [6–8]. After a week of exposure, almost zero weight loss was reported for a glassy Fe-based alloy in HCl, while austenitic stainless steels showed severe pitting under identical conditions [2]. The superior corrosion resistance of Fe-Cr-based amorphous alloys was attributed to the formation of a protective chromium oxyhydroxide passive layer on the surface, similar to stainless steel. However, due to their amorphous structure, Fe-Cr-based BMGs showed greater resistance to pitting corrosion as compared to stainless steel with similar Cr content [9]. A comparison between the corrosion rates of 304 stainless steel and Fe-10Cr-13P-7C amorphous alloy showed two to six-fold lower corrosion rates for the metallic glass in various concentrations of HCl. Similarly, weight loss measurements for $Fe_{67.7}B_{20}Cr_{12}Nb_{0.15}Mo_{0.15}$ BMG and stainless steel with similar Cr content in H_2SO_4 solution demonstrated 5–16 times lesser weight loss for the BMG [10].

4.1.1. Effect of Alloying Elements

Composition variation significantly influences the corrosion behavior of Fe-based BMGs. This effect may be classified into three groups [11]. The elements in the first group such as Cr and Ti help in the formation of a stable passive film and increase the corrosion resistance. The second group consists of elements that are more active than Fe including Vanadium (V), Niobium (Nb), Molybdenum (Mo), and Tungsten (W). These elements form corrosion product films covering the surface and act as diffusion barriers against further dissolution of the alloy. In contrast to the second group, the third group elements are nobler than Fe including Nickel (Ni), Cobalt (Co), Copper (Cu), Ruthenium (Ru), Rhodium (Rh), Palladium (Pd), and Platinum (Pt). The selective enrichment of these noble elements on the surface of the BMG in the corrosive environment has been shown to reduce the overall anodic activity and dissolution rate of Fe-based glassy alloys. The addition of different alloying elements (M) including Co, Ni, V, Cr, and Mn to Fe-M-$P_{13}C_7$ metallic glass, except for manganese (Mn), showed a reduction in the corrosion rate in HCl and H_2SO_4 solutions [12].

4.1.1.1. Effect of Chromium (Cr) Addition

Cr inhibits or hinders the dissolution of surface elements and helps in the formation of a passive protective film for a wide range of alloys. The high corrosion resistance of amorphous alloys containing chromium was attributed to the formation of the hydrated Cr oxyhydroxide film responsible for spontaneous passivation [11]. The passive film on Fe-Cr amorphous alloys is often composed of an outer Fe-rich layer and an inner Cr-rich layer. The Cr content in the alloy and the ratio of Cr/Fe determine the stability of the passive film. Further enhancement in corrosion resistance of the Fe-Cr glassy alloys was achieved by the addition of other metallic elements. For example, partial replacement of Fe by Al increased the corrosion resistance of $Fe_{45}Cr_{18}Mo_{14}C_{15}B_6Y_2$ in the presence of H_2. The corrosion rate of these alloys under conditions simulating the bipolar plate in the polymer electrolyte membrane fuel cell (PEMFC) follows the order: SUS316L (i_{corr} = 1.51 mA cm^{-2}) > $Fe_{43}Cr_{18}Mo_{14}C_{15}B_6Y_2Co_2$ (i_{corr} = 1.34) > $Fe_{45}Cr_{18}Mo_{14}C_{15}B_6Y_2$ (i_{corr} = 0.68) > $Fe_{43}Cr_{18}Mo_{14}C_{15}B_6Y_2Ni_2$ (i_{corr} = 0.57) > $Fe_{43}Cr_{18}Mo_{14}C_{15}B_6Y_2N_2$ (i_{corr} = 0.43) > $Fe_{43}Cr_{18}Mo_{14}C_{15}B_6Y_2Al_2$ (i_{corr} = 0.11) [13]. Figure 4.1 shows the decrease in corrosion rate of $([(Fe_{0.6}Co_{0.4})_{0.75}B_{0.2}Si_{0.065}]_{0.96}Nb_{0.04})_{100-x}Cr_x$ with increasing Cr content [14]. It was observed that the addition of Cr up to 4 at. % in this amorphous alloy system decreased the corrosion rate considerably in NaCl solution due to formation of a uniform protective passive layer. The addition of nitrogen had a more pronounced effect on the formation of the passive layer and corrosion resistance. Potentiodynamic studies for $Fe_{49}Cr_{15.3}Mo_{15}Y_2C_{15}B_{3.4}N_{0.3}$ revealed that the corrosion resistance of the N-containing amorphous alloy was at least one order of magnitude better than the N-free alloy in concentrated HCl solution [15]. The improved performance of the N-containing alloy was correlated to the presence of a thin MoN$^-$ layer over the protective passive film, which hindered the dissolution of Fe-oxide while preserving the Cr oxide. The Cl$^-$ ions in solution were repelled by the negatively charged N [16].

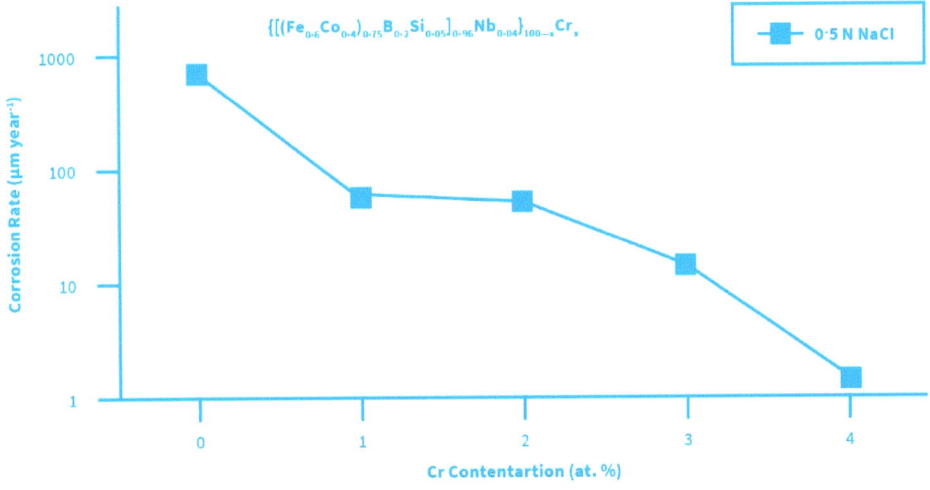

Figure 4.1. Effect of Cr addition on the corrosion rate of ([(Fe$_{0.6}$ Co$_{0.4}$)$_{0.75}$ B$_{0.2}$ Si$_{0.065}$]$_{0.96}$ Nb$_{0.04}$)$_{100-x}$ Cr$_x$ BMGs in 0.5 M NaCl at 298 K (data redrawn from reference [14]).

Several studies have shown that Fe-Cr-based amorphous alloys are more resistant to corrosion compared to commercially used alloys such as 316 stainless steel [17–19] and Ti-6Al-4V [20], and hence, are a potential substitute for conventional alloys in aggressive media [21]. Due to their excellent resistance to corrosion, they may be utilized as surgical tools or implants with more durability. Furthermore, because of the homogenous nature of the passive protective film, the Fe-Cr-based amorphous alloys require a lower content of Cr compared to their crystalline counterparts [22].

4.1.1.2. Effect of Molybdenum (Mo) Addition

The addition of Mo has been reported to have either a positive or negative effect depending on the alloy system. Alloying Fe-Cr-based BMGs with Mo prevents Cr dissolution and promotes passive layer formation. On the other hand, Mo dissolution has been observed even at lower potentials in the passive region and its oxide shows lesser stability compared to iron oxyhydroxide or hydrated chromium film. As a result, excessive addition of Mo to Fe-based amorphous alloys has been shown to deteriorate its electrochemical stability [23,24]. The effect of Cr and Mo addition on the passive current density of bulk glassy Fe$_{75-x-y}$Cr$_x$Mo$_y$C$_{15}$B$_{10}$ alloys in 1 N HCl has been evaluated by potentiodynamic polarization [25]. At fixed Mo concentration, the amorphous Fe$_{60-x}$Cr$_x$Mo$_{15}$C$_{15}$B$_{10}$ alloys containing 7.5–30 at. % Cr were shown to be instantly passivated with large passive regions until the transpassive dissolution of Cr. With the increase in Cr concentration, the anodic current density decreased significantly for the Fe$_{60-x}$Cr$_x$Mo$_{15}$C$_{15}$B$_{10}$ alloy system. At fixed Cr concentration, a higher fraction of Mo relative to Fe content was detrimental for corrosion resistance and reduced the protective characteristics of the passive film. Tungsten (W) had a similar beneficial influence at low concentration [26]. However, W addition was more effective than molybdenum in 6 M HCl solution because molybdenum exhibited an active dissolution peak in the cathodic region, while tungsten exhibited no active dissolution at potentials below 100 mV (vs. SCE) [27].

4.1.1.3. Effect of Other Metals

The addition of most transition metals to amorphous metal-metalloid alloys was found to improve their corrosion resistance [11]. A small addition of Ti, Mn, Nb, V, W, Ni, and Zr as partial substitutes for Fe in Fe-based metallic glasses decreased the corrosion rate [12,19,28,29]. Likewise, small addition of rare-earth elements such as Yttrium, Erbium (Er), and Dysprosium (Dy) to Fe-Cr-Mo-C-B helped increase corrosion resistance by the formation of a thick and highly stable passive film along with improvement in glass-forming ability (GFA) of the amorphous alloys [3,30,31]. Partial replacement of Fe by Co increased the corrosion resistance of $Fe_{80}B_{10}Si_{10}$ and $Fe_{75}P_{10}C_{10}B_5$ BMGs, resulting in nobler behavior and formation of protective film-containing Co oxides [32,33]. Manganese addition showed both positive and negative effects on corrosion resistance of different Fe-based BMGs. It deteriorated the corrosion resistance of $Fe_{80}P_{13}C_7$ glass, while improving the corrosion resistance of the $Fe_{44}Cr_{10}Mo_{12.5}Mn_{11}C_{15}B_6Y_{1.5}$ alloy system in HCl solution [34].

4.1.1.4. Effect of Metalloid Addition

The type and fraction of metalloids added to Fe-based BMGs were shown to affect the kinetics of passivation and composition of the passive film [30]. Addition of carbon (C) and boron (B) improved the corrosion resistance of Fe-BMGs [35,36] and enhanced their glass-forming ability. Increasing boron content (denoted by x) in glassy $Fe_{50-x}Cr_{16}Mo_{16}C_{18}B_x$ alloy system resulted in a decrease in the corrosion rate [30]. Figure 4.2 depicts the effect of B in glassy $Fe_{50-x}Cr_{16}Mo_{16}C_{18}B_x$ [14]. Increased boron content in this metallic glass system enhanced the corrosion resistance in various concentrations of HCl. Carbon was found to not influence the formation of passive film since carbonates are soluble in aqueous solutions, but it improved the corrosion resistance of Fe-based amorphous alloys in the presence of sufficient amounts of passivating elements [11].

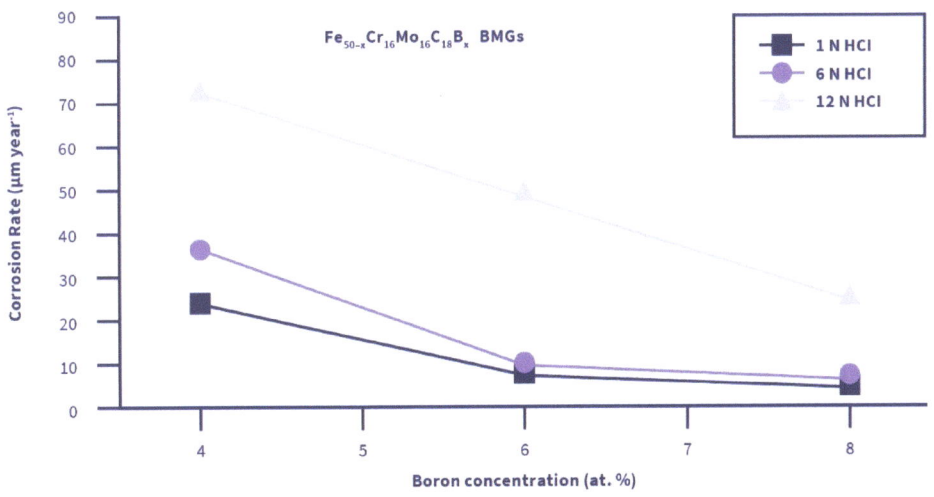

Figure 4.2. Corrosion rate of $Fe_{50-x}Cr_{16}Mo_{16}C_{18}B_x$ BMGs as a function of B content (denoted by x) in 1 N, 6 N, and 12 N HCl (data redrawn from Reference [14]).

The addition of Phosphorous (P) to Fe-based BMGs showed a positive effect in terms of corrosion resistance. Partial replacement of C and B by P in Fe-Cr-Mo-C-B glassy alloys resulted in the passivation current density (i_p) decreasing from ~10^{-1} to 5×10^{-2} A·m^{-2} in HCl [37]. Phosphorous enhances the passivity and stability of amorphous alloys in different ways. It accelerates the dissolution rate in the active range, leading to a rapid initial corrosion rate followed by passive element buildup at the surface [38,39]. Phosphorous generally forms a porous $FePO_4$ pre-passive film on the surface of Fe-based alloys, which facilitates the passivation and blocks actively dissolving sites. The passive Cr oxide layer then forms inside the pores of the pre-passive layer [40]. Additionally, P may be enriched underneath the passive film, acting as a chemical shield to reduce the ionic conductivity of the film/alloy interface, and consequently, inhibits anodic dissolution of the alloy and improves the stability of the passive film [38,41,42]. Silicon (Si) has been reported to be typically in the form of silicates in the surface film for Fe-based alloys but found to be not as beneficial in terms of accelerating passive film formation as phosphorus and carbon [11]. In summary, the beneficial effect of metalloid addition towards corrosion resistance improvement in Fe-based BMGs decreases in the order of phosphorus, carbon, silicon, and boron [11].

4.1.2. Effect of Structure and Crystallinity

The degree of short-range ordering and crystallinity has been reported to significantly affect the electrochemical behavior of Fe-based BMGs. As-cast $Fe_{73.5}Si_{13.5}B_9Nb_3Cu_1$ and $Fe_{48}Cr_{15}Mo_{14}C_{15}B_6Y_2$ BMGs showed the best resistance to corrosion compared with their annealed nanocrystalline and fully crystallized counterparts in HCl and H_2SO_4 [43,44]. This was attributed to the chemical heterogeneity (galvanic coupling) due to the formation of precipitates, segregation, and other composition fluctuations leading to local galvanic cells. Similar behavior was reported for other Fe-based amorphous alloys [36,45,46]. The corrosion current density was lower for $(Fe_{44.3}Cr_5Co_5Mo_{12.8}Mn_{11.2}C_{15.8}B_{5.9})_{98.8}Y_{1.5}$ glassy alloys compared to X210Cr12 steel in 0.5 M H_2SO_4 [47]. Due to galvanic coupling between the matrix and inter-dendritic phase(s) in the crystalline counterpart, it showed lower corrosion resistance as compared with the fully amorphous alloy. However, certain Fe-based BMGs showed better corrosion resistance after crystallization in comparison with the fully amorphous as-cast state due to the passive nature of the corrosion products on the surface of the alloys [48].

Recently, Fe-based amorphous coatings have been studied for surface engineering applications. The effect of partial crystallization on the corrosion behavior of $Fe_{48}Cr_{15}Mo_{14}C_{15}Y_2B_6$ amorphous coating fabricated by high-velocity oxygen fuel (HVOF) was studied, showing that the increase in crystalline fraction lowered its corrosion resistance [49,50]. The detrimental effect of crystallization on the corrosion resistance of these alloys was attributed to carbide precipitation during heat treatment, which resulted in nanoscale Cr-depleted zones [44,51]. SEM and TEM micrographs showed that corrosion pits initiated at the boundaries around the inter-splats where Cr and Mo were depleted [50]. X-ray photoelectron spectroscopy (XPS) on the passive film confirmed a significant decrease in Cr oxides and Mo oxides in the crystallized coating. This reduction accounted for the deterioration of electrochemical resistance in the crystallized coating since Cr and Mo oxides are more protective than Fe oxides. In other words, a large portion of Cr and Mo was consumed as a result of Cr-/Mo-rich carbide precipitation in the crystallized coating, resulting in the deterioration of corrosion resistance. Similar behavior has been reported for other Fe-based amorphous alloys such as Fe-Cr-P-C and Fe-Ni-Cr-P-B [46].

4.1.3. Effect of Test Environment

Fe-based BMGs showed excellent corrosion resistance in various aggressive media such as HCl, H_2SO_4, NaCl, Na_2SO_4, NaOH, and simulated body fluid (SBF) [13,15,20,21]. The sodium chloride (NaCl) environment has been widely studied for several Fe-based metallic glasses and coatings. Due to the presence of Cl^- ions in the electrolyte, chloride-induced pitting was commonly reported. The difference (ΔE) between corrosion potential (E_{corr}) and pitting potential (E_p) was determined using accelerated polarization experiments. Very noble levels for pitting onset potential were shown for $Fe_{56}Co_7M_2Mo_5Zr_{10}B_{20}$ (M = W or Ni) [52] in 3.5% NaCl with higher ΔE for the W-containing alloy (852 mV vs. SCE) compared to the Ni-containing alloy (349 mV vs. SCE). Electrochemical impedance spectroscopy (EIS) plots for the

BMGs showed a single capacitance loop and the polarization resistance (R_p) for the W-containing alloy was higher than the Ni-containing alloy. Quaternary Fe-Ni-P-C, Fe-Ni-P-B, and Fe-Cr-P-C BMGs also showed superior corrosion resistance in NaCl solution [13,53,54]. The concentration of chloride ions in solution showed a significant impact on the corrosion behavior. Although potentiodynamic polarization plots for $Fe_{41}Co_{17}Cr_{17}Mo_{14}C_{15}B_6Y_2$ obtained in different concentrations of NaCl solution showed a similar passivation profile, increased solution concentration resulted in a decrease in corrosion potential (E_{corr}) and increase in corrosion current density (i_{corr}) [55].

It was recently shown that two amorphous alloys, namely $Fe_{48}Mo_{14}Cr_{15}Y_2C_{15}B_6$ (SAM1651) and $Fe_{49.7}Cr_{17.7}Mn_{1.9}Mo_{7.4}W_{1.6}B_{15.2}C_{3.8}Si_{2.4}$ (SAM2X5), demonstrated the highest corrosion resistance in 3.5% NaCl among Fe-based amorphous coatings [17,50]. These coatings were fabricated by the HVOF (high-velocity oxygen fuel) of glassy powders and were characterized by spontaneous passivation with a wide passive region and low passive current density. The coating with coarse powders (45–55 μm) exhibited better corrosion resistance than the one with fine powders (20–33 μm), even though it contained more porosity. This behavior was attributed to more oxidation and oxide-rich inter-splats in finer powder coatings as compared to the one with coarse powder. Fe-Cr-Mn-Mo-W-B-C-Si amorphous coatings fabricated by HVOF and HVAF (high-velocity air fuel) processes showed different electrochemical behaviors in NaCl solution [56]. More oxides formed during HVOF hindered the formation of dense passive film and reduced the corrosion resistance. The corrosion resistance of Fe-based BMGs in NaCl solution was reported to be better than 304 stainless steel and was typically accompanied by the formation of a stable passive layer with low passive current density [57].

Numerous investigations have been carried out on the corrosion and pitting resistance of Fe-based BMGs in HCl, which is the typical environment in applications such as pickling, ore reduction, food processing, and metal cleaning. The effects of acid concentration and temperature were evaluated for different BMGs. Fe-based BMGs showed spontaneous passivation and a wide passive region [17,26–29,53,58,59]. However, with increasing concentration of the corrosive medium, the corrosion rate increased [30,60]. Moreover, increasing temperature increased the rate of the anodic process with the subsequent localized attack on the alloy surface. Fe-based BMGs in the HCl environment generally showed better corrosion resistance as compared with stainless steel.

Sulfuric acid (H_2SO_4) is another electrolyte in which the corrosion behavior of Fe-based BMGs has been evaluated extensively because sulfuric acid is used in a wide range of applications including pickling, petroleum industries, batteries, fuel cells, and the pulp and paper industry [13,33,61,62]. Fe-Cr-P-C [12], $Fe_{75}P_{10}C_{10}B_5$ [33], and $Fe_{73.5}Si_{13.5}B_9Nb_3Cu_1$ [48] BMGs demonstrated excellent corrosion resistance in H_2SO_4. The addition of elements such as Mo, Ni, Cr, and Co to Fe-based BMGs resulted in further enhancement of their corrosion resistance [33]. Increasing acid concentration was found to decrease E_{corr} towards less noble values and increase i_{corr} [63]. A recent study on the corrosion behavior of $Fe_{48}Cr_{15}Mo_{14}C_{15}B_6Y_2$ amorphous coating fabricated by HVOF demonstrated that unlike the behavior in HCl and NaCl solutions, the corrosion resistance of amorphous coatings in H_2SO_4 was inferior compared to 316L stainless steel. This behavior was attributed to the aggressive oxidizing environment of H_2SO_4 as compared to HCl and easier passivation of stainless steel in H_2SO_4 [17]. The corrosion behavior of Fe-based metallic glasses was evaluated in the highly aggressive $CaCl_2$ solution at high temperatures. Amorphous $Fe_{58.4}Cr_{18.5}B_{3.2}Mo_{14}C_1Si_{1.3}Mn_2W_{5.8}$ showed uniform corrosion in 5 M $CaCl_2$ while stainless steels and Ni-based alloys showed severe pitting [64].

Some BMGs have been reported to have good biocompatibility in addition to good corrosion resistance that makes them potentially promising in biomedical applications. Most biocompatibility studies have been performed for Zr-based and Ti-based BMGs and there are fewer reports on Fe-based BMGs [58,65,66]. Biocompatibility evaluation of Fe-based BMGs has been conducted in terms of ion release in simulated body fluids and cytotoxicity assessment. Electrochemical studies for Fe-based BMGs such as $Fe_{41}Co_7Cr_{15}Mo_{14}C_{15}B_6Y_2$, $Fe_{44}Cr_5Co_5Mo_{13}Mn_{11}C_{16}B_6Y_2$, and $Fe_{48}Cr_{15}Mo_{14}C_{15}B_6Er_2$ in Hank's solution and artificial saliva showed their superior corrosion performance as compared with SS316 L [67,68]. These BMGs also showed no cytotoxicity to L929 and NIH3T3 cells. Co and Mn may induce adverse body reactions in long-term implantation [69]. Therefore, new Fe-based metallic glasses with higher Fe content and free from Co and Mn have recently been studied for implant applications [70]. These BMGs demonstrated better corrosion resistance than biomedical grade alloys such as SS316 and Ti-6Al-4V [70]. In electrochemical studies for a series of $Fe_{80-x-y}Cr_xMo_yP_{13}C_7$ BMGs in simulated body fluids (Hank's solution and artificial saliva), the amorphous alloys showed high corrosion resistance and the released ion content (measured by inductively coupled plasma atomic emission spectroscopy (ICP-AES)) was lower than SS316L. The addition of Nb further enhanced the corrosion resistance of $Fe_{55-x}Cr_{18}Mo_7B_{16}C_4Nb_x$ in Ringer's solution even more than conventional 316L stainless steel and Ti-6Al-4V biomedical alloys, with no ions released in the solution [19].

Weight loss measurements have been reported for the $(Fe_{44.3}Cr_5Co_5Mo_{12.8}Mn_{11.2}C_{15.8}B_{5.9})_{98.5}Y_{1.5}$ alloy in four different solutions including very acidic (0.5 M H_2SO_4), very basic (1 M NaOH), neutral aggressive chloride solution (0.6 M NaCl), and less aggressive neutral solution (0.1 M Na_2SO_4). These studies indicated that Fe-based glassy alloys had very low weight-loss and high corrosion resistance in NaCl, Na_2SO_4, and NaOH solutions but relatively less corrosion resistance in H_2SO_4. SEM images of the surface of $(Fe_{44.3}Cr_5Co_5Mo_{12.8}Mn_{11.2}C_{15.8}B_{5.9})_{98.5}Y_{1.5}$ BMG after immersion in H_2SO_4 and NaOH for 100 days showed that the surface remained mostly intact in NaOH [4]. In addition, it was shown that at higher pH values (in NaOH), the passivation ability of the amorphous and crystalline $(Fe_{44.3}Cr_5Co_5Mo_{12.8}Mn_{11.2}C_{15.8}B_{5.9})_{98.5}Y_{1.5}$ alloys was improved [4], demonstrating that galvanic corrosion was less pronounced in alkaline media [47]. The corrosion current density (i_{corr}), corrosion potential (E_{corr}), and pitting potential (E_{pit}) of Fe-based BMGs as a function of environment and test temperature are summarized in Table 4.1 below.

Table 4.1. Corrosion parameters for Fe-based BMGs in different environments.

Fe-Based BMG	i_{corr} ($\mu A/cm^2$)	E_{corr} (mV vs. SCE)	E_{pit} (mV vs. SCE)	Environment	T (°C)	Ref.
Simulated Body Fluids						
$Fe_{60}Cr_{10}Mo_{10}P_{13}C_7$	0.45	−317	977	Hank's	25	[70]
$Fe_{60}Cr_{10}Mo_{10}P_{13}C_7$	0.22	−314	1123	Saliva	25	[70]
$Fe_{55}Cr_{18}Mo_7B_{16}C_4$	2.46	−169	715	Ringer	37	[19]
$Fe_{52}Cr_{18}Mo_7B_{16}C_4Nb_3$	0.19	−45	876	Ringer	37	[19]
$Fe_{51}Cr_{18}Mo_7B_{16}C_4Nb_4$	0.03	122	1299	Ringer	37	[19]
$Fe_{41}Co_7Cr_{15}Mo_{14}C_{15}B_6Y_2$	0.23	−270	-	Hank's	36	[68]
$Fe_{41}Co_7Cr_{15}Mo_{14}C_{15}B_6Y_2$	0.048	−320	-	Saliva	36	[68]
$(Fe_{44}Cr_5Co_5Mo_{13}Mn_{11}C_{16}B_6)_{98}Y_2$	0.28	−230	-	Hank's	36	[68]
$(Fe_{44}Cr_5Co_5Mo_{13}Mn_{11}C_{16}B_6)_{98}Y_2$	0.07	−230	-	Saliva	36	[68]
$Fe_{48}Cr_{15}Mo_{14}C_{15}B_6Er_2$	0.09	−300	-	Hank's	36	[68]
$Fe_{48}Cr_{15}Mo_{14}C_{15}B_6Er_2$	0.003	−250	-	Saliva	36	[68]
$Fe_{41}Co_7Cr_{15}Mo_{14}C_{15}B_6Y_2$	0.23	−269	1009	Hank's	25	[67]
$Fe_{41}Co_7Cr_{15}Mo_{14}C_{15}B_6Y_2$ (Annealed)	0.35	−286	570	Hank's	25	[67]
$Fe_{41}Co_7Cr_{15}Mo_{14}C_{15}B_6Y_2$	0.04	−315	1200	Saliva	25	[67]
$Fe_{41}Co_7Cr_{15}Mo_{14}C_{15}B_6Y_2$ (Annealed)	0.14	−290	310	Saliva	25	[67]
$Fe_{55}Cr_{20}Mo_5P_{13}C_7$	0.26	−306	1040	Hank's	25	[70]
$Fe_{55}Cr_{20}Mo_5P_{13}C_7$	0.17	−352	1123	Saliva	25	[70]
$Fe_{50}Cr_{20}Mo_{10}P_{13}C_7$	0.37	−309	995	Hank's	25	[70]
$Fe_{50}Cr_{20}Mo_{10}P_{13}C_7$	0.20	−364	1099	Saliva	25	[70]
NaCl						
$Fe_{60}Ni_{20}P_{14}B_6$	10.96	−364	-	0.5 M NaCl	25	[53]
$Fe_{50}Ni_{30}P_{14}B_6$	7.02	−343	-	0.5 M NaCl	25	[53]
$Fe_{48}Cr_{15}Mo_{14}C_{15}Y_2B_6$ (coating)	0.99	−304	-	0.6 M NaCl	25	[49]
Annealed at 530 °C for 5 h	1.88	−402	-	0.6 M NaCl	25	[49]
Annealed at 600 °C for 1 h	4.97	−455	-	0.6 M NaCl	25	[49]
Annealed at 700 °C for 1 h	14.7	−592	-	0.6 M NaCl	25	[49]
$Fe_{48}Mo_{14}Cr_{15}Y_2C_{15}B_6$ (coating)	6.1	−446	1354	0.6 M NaCl	25	[50]
HCl						
$Fe_{66.7}C_{7.0}Si_{3.3}B_{5.5}P_{8.7}Cr_{2.3}Al_{2.0}Mo_{4.5}$	4.06	−283		1 N HCl	25	[45]
Annealed at 783 K for: 1 h	4.34	−310		1 N HCl	25	[45]
Annealed at 783 K for 4 h	5.06	−310		1 N HCl	25	[45]
Annealed at 783 K for 24 h	5.67	−312		1 N HCl	25	[45]
$Fe_{69.9}C_{7.1}Si_{3.3}B_{5.5}P_{8.7}Mo_{2.5}Al_{2.0}Co_{1.0}$	4.5	−311	-	1 N HCl	25	[58]
$Fe_{67.6}C_{7.1}Si_{3.3}B_{5.5}P_{8.7}Cr_{2.3}Mo_{2.5}Al_{2.0}Co_{1.0}$	1.9	−290	-	1 N HCl	25	[58]
$Fe_{60}Ni_{20}P_{14}B_6$	58.91	−176	-	1 M HCl	25	[53]
$Fe_{50}Ni_{30}P_{14}B_6$	41.76	−166	-	1 M HCl	25	[53]
H_2SO_4						
$Fe_{69.9}C_{7.1}Si_{3.3}B_{5.5}P_{8.7}Mo_{2.5}Al_{2.0}Co_{1.0}$	6.5	−304	-	0.5 M H_2SO_4	25	[58]
$Fe_{67.6}C_{7.1}Si_{3.3}B_{5.5}P_{8.7}Cr_{2.3}Mo_{2.5}Al_{2.0}Co_{1.0}$	2.3	−279	-	0.5 M H_2SO_4	25	[58]
$Fe_{66.7}C_{7.0}Si_{3.3}B_{5.5}P_{8.7}Cr_{2.3}Al_{2.0}Mo_{4.5}$	0.9	−264		0.5 M H_2SO_4	25	[45]
Annealed at 783 K for: 1 h	2.1	−298		0.5 M H_2SO_4	25	[45]
Annealed at 783 K for:4 h	2.50	−302		0.5 M H_2SO_4	25	[45]
Annealed at 783 K for:24 h	4.29	−312		0.5 M H_2SO_4	25	[45]
$Fe_{57.6}C_{7.1}Si_{3.3}B_{5.5}P_{8.7}Cr_{12.3}Mo_{2.5}Al_{2.0}Co_{1.0}$	0.7	−235	-	0.5 M H_2SO_4	25	[58]

Table 4.1. Cont.

Fe-Based BMG	i_{corr} (μA/cm²)	E_{corr} (mV vs. SCE)	E_{pit} (mV vs. SCE)	Environment	T (°C)	Ref.
$Fe_{41}Co_7Cr_{15}Mo_{14}C_{15}B_6Y_2$ Other solutions	2.42	−196	-	1 M H_2SO_4	25	[63]
$Fe_{43}Cr_{18}Mo_{14}C_{15}B_6Y_2Co_2$	1.34×10^3	-	-	1 M H_2SO_4 + 2 ppm F^- (H_2 bubbling)	80	[13]
$Fe_{45}Cr_{18}Mo_{14}C_{15}B_6Y_2$	0.68×10^3	-	-	1 M H_2SO_4 + 2 ppm F^- (H_2 bubbling)	80	[13]
$Fe_{43}Cr_{18}Mo_{14}C_{15}B_6Y_2Ni_2$	0.57×10^3	-	-	1 M H_2SO_4 + 2 ppm F^- (H_2 bubbling)	80	[13]
$Fe_{43}Cr_{18}Mo_{14}C_{15}B_6Y_2N_2$	0.43×10^3	-	-	1 M H_2SO_4 + 2 ppm F^- (H_2 bubbling)	80	[13]
$Fe_{43}Cr_{18}Mo_{14}C_{15}B_6Y_2Al_2$	0.11×10^3	-	-	1 M H_2SO_4 + 2 ppm F^- (H_2 bubbling)	80	[13]
$Fe_{43.2}Co_{28.8}B_{19.2}Si_{4.8}Nb_4$	15	-	-	0.5 M Na_2SO_4	25	[66]
$Fe_{41.47}Co_{27.65}B_{18.43}Si_{4.61}Nb_{3.84}Cr_4$	0.67	-	-	0.5 M Na_2SO_4	25	[66]

4.2. Ni-Based Metallic Glasses

Ni-based BMGs have drawn wide attention in the last few years due to their high strength and toughness, high corrosion resistance, good glass-forming ability, and thermal stability [71,72]. Ni-based amorphous alloys showed higher corrosion resistance than stainless steels under conditions simulating proton exchange membrane fuel cells (PEMFCs) [73]. A variety of Ni-based BMGs with excellent corrosion resistance have been developed to date including Ni-Cr-P-B(Mo), Ni-Cr-Ta-(Nb)-P-B, Ni-Nb-Ta-P, Ni-Cr-Ta-Mo-P-B, Ni–Ti–Zr–Si–Sn, Ni–Nb–Ti–Zr–Co–Cu, Ni–Cu–Ti–Zr–Al, Ni–Ta–(Nb)–Sn, among others. These alloys passivated spontaneously and did not show pitting corrosion even in highly concentrated solutions [74–78]. Ni-based amorphous alloys such as Ni-Zr-Ti-Si-Sn or Ni-Nb-Ti-Zr-Co-Cu were found to be good candidates as coatings in aggressive environments for anticorrosion applications [79,80]. In addition, thicker coatings showed higher corrosion resistance because of a lower number of through-pores with increasing thickness [80]. In addition to chemical composition, the corrosion behavior of Ni-based amorphous alloys was found to be dependent on their structural homogeneity.

4.2.1. Effects of Alloying Elements

Addition of Cr and Mo in different proportions resulted in a wide range of corrosion behaviors [74,81,82]. The $Ni_{77-x-y}Mo_xCr_yNb_3P_{14}B_6$ bulk metallic glass system exhibited highest corrosion resistance at x = 5 and y = 8, with a very low corrosion current density and a corrosion rate less than 10^{-3} mm/year in both HCl and NaCl (Table 4.2) [71]. Tungsten addition showed an affect similar to that of molybdenum [83].

Table 4.2. EIS results for $Ni_{77-x-y}Mo_xCr_yNb_3P_{14}B_6$ BMGs in 1 M HCl (R_s: solution resistance; R_p: polarization resistance; i_{corr}: corrosion current density; R_{corr}: corrosion rate) (data taken from reference [71]).

$Ni_{77-x-y}Mo_xCr_yNb_3P_{14}B_6$	x, y (at. %)	R_s (Ω cm^2)	R_p (×10^4 Ω cm^2)	i_{corr} (10^{-6} A cm^{-2})	R_{corr} (mm/year)
	x = 7 y = 0	0.106	0.854	7.38	0.057
	x = 8 y = 0	0.119	1.180	3.49	0.026
	x = 9 y = 0	0.102	0.745	9.91	0.078
	x = 5 y = 3	0.111	1.01	3.35	0.024
1 M HCl solution	x = 5 y = 5	0.135	1.98	1.34	<10^{-3}
	x = 5 y = 8	0.122	2.38	1.10	<10^{-3}
	x = 8 y = 3	0.127	1.51	2.50	0.019
	x = 8 y = 5	0.103	1.66	2.03	0.015

The addition of Cr, Mo, and Ta together was found to be very effective in achieving high corrosion resistance. The bi-layer passive film consisted of outer stable triple oxyhydroxide, $Cr_{1-x-y}Ta_xNb_yO_z(OH)_{3+2x+2y-2z}$, and an inner MoO_2 layer acted as a diffusion barrier promoting high corrosion resistance [74,83,84]. Palladium addition has also shown a positive effect in terms of improving the corrosion resistance of Ni-based BMGs [85]. Niobium and Tantalum were found to form Nb- and Ta-enriched passive layers in Ni-based BMGs [78,86–90], with Ta addition resulting in better corrosion resistance compared to Nb addition. Nb, Ti, and Zr were preferentially oxidized to form enriched films in Ni-Nb-Ti-Zr-Cu BMGs when exposed to air. Nb surface enrichment increased due to the easier dissolution of Ni when the alloy was immersed in acidic electrolytes (HCl and H_2SO_4) [91]. XPS analysis revealed that with increasing Zr in Ni-Nb-Zr, Nb- and Zr-enriched surface films formed and hindered the dissolution reactions [75]. Phosphorous was very effective among metalloids in reducing the corrosion rate of Ni-based amorphous alloys as it acted as a diffusion barrier [92,93].

4.2.2. Effects of Structure and Crystallinity

In addition to the chemical composition of Ni-based BMGs, their microstructure has been reported to affect their corrosion behavior. The corrosion resistance of Ni-based BMGs was better than their devitrified (crystallized) counterparts due to the chemical and structural homogeneity as reported for Ni-10Ta-20P and $Ni_{55}Nb_{20}Ti_{10}Zr_8Co_7$ alloys [94–96]. However, some partially crystallized amorphous alloys exhibited better corrosion resistance. Partial devitrification in $Ni_{55}Nb_{30}Sn_5Ti_5Zr_5$ and $Ni_{55}Nb_{30}Sn_5Zr_{10}$ BMGs showed superior corrosion resistance compared to their fully amorphous counterparts in NaCl solution [72]. In addition, $Ni_{62}Nb_{33}Zr_5$ BMG showed inferior pitting resistance than its partially crystallized counterpart in HCl [97]. The electrochemical behavior for both $Ni_{59}Zr_{20}Ti_{16}Si_2Sn_3$ and $Ni_{53}Nb_{20}Ti_{10}Zr_8Co_6Cu_3$ alloys were similar in crystalline and amorphous states, indicating that homogenous amorphous structure may not always result in superior corrosion properties [72,78]. This behavior was attributed to the faster formation of a protective passive layer due to the accelerated diffusion of passive film formers via the amorphous/crystalline interface [98,99].

4.2.3. Effects of Test Environment

Ni-based amorphous alloys have been widely studied in aggressive environments where the corrosion resistance was found to depend on solution composition, temperature, and pH. Ni-Ta-(P,Sn), Ni-Nb-Zr, Ni-Nb-Ta-P, Ni-Cr-Nb-P-B, Ni-Zr-Ti-Si-Sn, Ni-Cr-Mo-Ta-Nb-P, and Ni-Nb-Ti-Zr-Co-Cu metallic glasses showed good corrosion resistance in different concentrations of HCl (1 to 12 M HCl) [74–76,80, 82,83,90,92,100–102]. The corrosion behavior of Ni-based amorphous alloys has also been well studied in H_2SO_4 solution [73,87,95,103]. Ni-based BMGs showed excellent corrosion resistance in aggressive 1 M H_2SO_4 + 2ppm F^- at 80 °C as well as NaOH solutions [104,105]. Overall, Ni-based BMGs showed good corrosion resistance in acidic, chloride ions-containing neutral, and alkaline solutions, which makes them good candidates for wide ranging structural applications.

4.3. Cobalt (Co)-Based Metallic Glasses

Co-based metallic glasses are known for their soft magnetic properties and good mechanical behavior including high fracture toughness [106]. $Co_{43}Fe_{20}Ta_{5.5}B_{31.5}$ metallic glass exhibited high fracture strength of about 5.3 GPa [107]. There are few reports on the corrosion behavior of Co-based amorphous alloys. $Co_{43}Fe_{20}Ta_{5.5}B_{31.5}$ BMG showed very good corrosion resistance in NaCl (corrosion rate of 5.6×10^{-3} mm/year), H_2SO_4 (8.3×10^{-3} mm/year), and HCl (1.3×10^{-2} mm/year) with no pitting in Cl^--containing solutions as well as excellent passivation behavior [108]. The replacement of Fe by B in this system and the addition of Si further enhanced the corrosion resistance. Lower corrosion rates in HCl and NaCl have been reported for both $Co_{62.2}B_{26.9}Si_{6.9}Ta_4$ and $Co_{65.9}B_{24.7}Si_{5.4}Ta_4$ BMGs, with wide passive ranges [109]. The addition of 8 at. % Cr to $Co_{73.5}Si_{13.5}B_9Nb_3Cu_1$ metallic glass shifted the polarization curves towards lower i_{corr} and nobler E_{corr} in H_2SO_4 and Na_2SO_4 solutions, resulting in better corrosion resistance compared with Fe-based BMGs [48]. However, the corrosion resistance was still relatively low, suggesting that the amount of Cr was not enough to create a stable passive film in these solutions [48,110]. The addition of Mo and W increased the corrosion stability of Co-based metallic glasses as well [111]. Although most investigations on Co-based BMGs were focused on alloy development, a recent study has reported their potential for biomedical applications. $Co_{80-x-y}Cr_xMo_yP_{14}B_6$ (x = 5, y = 5; x = 5, y = 10; x = 10, y = 10) BMGs exhibited good corrosion resistance in Hank's solution and artificial saliva for x=y=10. There was an insignificant amount of ions released and higher cell viability in these solutions as compared with SS316L and CoCrMo biomedical alloys, indicating good biocompatibility for dental implantation [112]. Furthermore, Co-based amorphous alloys exhibited good corrosion resistance in phosphate-buffered saline (PBS) solution, making them suitable for biosensors [113].

The electrochemical properties of Co-based metallic glasses have been reported to be dependent on their thermal history. $Co_{73.5}Si_{13.5}B_9Nb_3Cu_1$ amorphous and nanocrystalline alloys displayed better corrosion resistance at lower acid concentrations (1 M H_2SO_4) compared to their polycrystalline counterparts. However, the polycrystalline alloys exhibited better corrosion behavior at a higher concentration of H_2SO_4. The better immunity of crystalline alloys to corrosion in highly acidic environments was attributed to the formation of a surface layer of corrosion products, which acted as an effective barrier layer between the material and the aggressive environment [48]. Following a similar mechanism, heat-treated Co-Si-B amorphous alloys demonstrated very good electrochemical characteristics in NaCl solution, which was attributed to the higher diffusion of Si to the surface and formation of the protective SiO_2 film [114]. Different concentrations of the electrolyte affected the electrochemical behavior of Co-based metallic glasses in different ways. The corrosion rate of $Co_{73.5-x}Si_{13.5}B_9Nb_3Cu_1Cr_x$ (x=0 and 8) metallic glass decreased with increasing H_2SO_4 concentration (1, 3, and 5 M) [48].

4.4. Copper (Cu)-Based Metallic Glasses

Cu-based BMGs are typically known for their relatively low cost, high glass-forming ability, good thermal stability, and relatively good corrosion properties [115]. Since the production cost for these BMGs is relatively low and due to their good ductility and strength, Cu-Zr-based BMGs are attractive for biomedical applications, microdevices, and bipolar plates compared to metallic glasses based on toxic elements such as Be or Ni [116]. However, the formation of highly localized shear bands and shear softening are some critical issues specifically in Cu-based BMGs that result in catastrophic failure. One of the main reasons for limited utilization of Cu-based BMGs is inhomogeneous deformation and brittle failure. In order to overcome this challenge, some attempts have been made to develop reinforced BMGs via various routes including partial crystallization, particle reinforcement, and in situ precipitation [117]. Consequently, most of the corrosion investigations are focused on these composite microstructure-based Cu-BMGs.

4.4.1. Effect of Alloying Elements

The effects of different alloying elements on the corrosion behavior of Cu-based BMGs have been systematically investigated. The addition of 10 at. % Nb to Cu-Hf-Ti BMG resulted in a composite microstructure and reduced the corrosion rate of the alloys. Through cationic concentration analysis in the passive layer, this improvement was attributed to the formation of Nb-, Hf-, and Ti-enriched protective surface films in NaCl and H_2SO_4 + NaCl solutions [117]. Niobium (Nb) addition promoted the corrosion resistance of $Cu_{50-x}Zr_{45}Al_5Nb_x$ (x = 0, 1, 3, and 5 at. %) [118] and $Cu_{55-x}Zr_{40}Al_5Nb_x$ (x=0–5 at. %) [119] metallic glasses in NaCl and HCl by decreasing Cu concentration in the passive film while increasing the fraction of passivating elements [118,120]. Nb addition to other Cu-based BMGs including $(Cu_{0.36}Zr_{0.48}Ag_{0.08}Al_{0.08})_{95}Nb_5$ [121], $(Cu_{0.6}Zr_{0.3}Ti_{0.1})_{100-x}Nb_x$ (x = 0–5 at %) [120,122], and $(Cu_{0.6}Hf_{0.25}Ti_{0.15})_{98}Nb_2$ [123] showed the same beneficial effect in terms of corrosion resistance in HCl, NaCl, H_2SO_4, HNO_3, and NaOH electrolytes, with the enrichment of Zr, Ti, and Nb and lowering of Cu content in the passive film [121]. Simultaneous addition of Ni and Nb has been reported to be very effective in reducing pitting susceptibility and improvement of pitting corrosion resistance for Cu-Zr-Ti and Cu–Hf–Ti BMGs in different solutions [124–127].

Another effective passivity promoter is Ti which has been added to many metallic glass systems to improve their corrosion resistance. Ti micro-alloying in a Cu-Zr-Ag-Al-Ti BMG resulted in lowering of the passive current density and increase in the corrosion potential in both H_2SO_4 and NaOH [116]. The corrosion resistance of $Cu_{60}Zr_{30}Ti_{10}$ BMG, which contains two passive elements (Zr and Ti), was reported to be relatively high in H_2SO_4, HNO_3, NaOH, and NaCl electrolytes [128]. Addition of Mo led to the enhancement in corrosion resistance for $Cu_{60}Zr_{30}Ti_{10}$ [122], $Cu_{60}Hf_{25}Ti_{15}$ [123], and $Cu_{47}Zr_{11}Ti_{34}Ni_8$ BMGs [129,130] in various electrolytes including HCl, HNO_3, NaOH, and NaCl. Mo suppressed the diffusion of Cu or Ni to the passive film and formed a Zr-/Ti-enriched protective film [129,131] that was structurally denser and more protective than the Cu-/Ni-oxide film. Mo was present in the form of MoO_2 in the inner layer of the passive film and stabilized the surface layer [129]. Additionally, Mo changed the shape of potential–time curves in both acidic and alkaline solutions and accelerated the formation of a more stabilized passive film. Small additions of Ta [122,123], W [130], and Cr [130] along with some rare-earth elements such as Y [132], indium (In) [133], lanthanum (La) [134], and cerium (Ce) [135] were also reported to enhance the corrosion resistance of Cu-based BMGs.

The influence of partial and complete devitrification of Cu-based metallic glasses on their corrosion behavior has been studied extensively. Partial crystallization of $Cu_{50}Zr_{45}Al_5$ BMG led to enhancement in corrosion resistance in different solutions owing to the higher diffusion rate of the passive elements for faster formation of protective surface film in the presence of amorphous/crystalline interfaces. However, there was rapid degradation in corrosion resistance for the fully crystallized state [98]. In addition, the $Cu_{47.5}Zr_{47.5}Al_5$ BMG composite demonstrated superior corrosion resistance in comparison to monolithic glass and annealed composite in seawater solution due to the presence of nanocrystalline CuZr particles distributed homogeneously in the glassy matrix which accelerated the passive film formation [115]. By a similar mechanism, the $(Cu_{0.6}Hf_{0.25}Ti_{0.15})_{90}Nb_{10}$ composite showed excellent corrosion resistance in acid- and chloride ion-containing solutions [117].

4.4.2. Effect of Test Environment

Cu-based BMGs have been reported to be susceptible to pitting corrosion in chloride-containing solutions, while their corrosion resistance in H_2SO_4 and HNO_3 was reported to be relatively better [122,123,125,133,136]. The influence of Cl^- concentration on the corrosion behavior of Cu-based metallic glasses has been widely reported. The passive region of $Cu_{46}Zr_{42}Al_7Y_5$ BMG in NaCl became smaller with increasing chloride concentration along with an increase in pitting susceptibility [137–139]. Similar results were reported for the corrosion behavior of $Cu_{60}Zr_{20}Ti_{20}$ [140], $Cu_{60}Zr_{30}Ti_{10}$ [138], and $Cu_{55}Zr_{35}Ti_{10}$ [139] glassy alloys in both HCl and NaCl. This was attributed to the lower oxygen content in solution with increasing chloride concentration resulting in a decrease in the thickness of the protective oxide passive film [139].

4.5. Chromium (Cr)-Based Metallic Glasses

Chromium (Cr) is an excellent passivity promoter and Cr-containing BMGs have been widely studied in terms of their corrosion behavior in different environments. Similar to the effect in crystalline alloys, Cr addition in metallic glass systems has been reported to improve their corrosion resistance by stabilization of the protective passive film. In the previous sections, the addition of Cr to different Fe-, Ni-, Co-, and Cu-based BMGs was shown to improve their corrosion behavior. However, there are limited reports on Cr-based BMGs. Increasing Cr content in Cr-Fe-Mo-C-B-Y BMGs resulted in an increase in the open circuit potential (OCP) in HCl, indicating high corrosion resistance in immersion tests [141]. However, there was a threshold in terms of Cr content above which no further improvement in corrosion resistance was observed. The electrochemical behavior of $Cr_{40}Co_{39}Nb_7B_{14}$ and $Cr_{50}Co_{29}Nb_7B_{14}$ systems was evaluated in HCl and no weight loss was observed for both the alloys even after a long duration due to the resistivity of the surface passive film [142]. These limited studies show that there is great potential in terms of alloy development for Cr-based BMGs as well as characterization of their corrosion behavior in different environments.

References

1. Naka, M.; Hashimoto; Masumoto, T. Corrosion resistivity of amorphous Fe alloys containing Cr. *J. Jpn Inst. Met.* **1974**, *38*, 835–841. (In Japanese) [CrossRef]

2. Naka, M.; Hashimoto, K.; Masumoto, T. High corrosion resistance of Cr-bearing amorphous Fe alloys in neutral and acidic solutions containing chloride. *Corrosion* **1976**, *32*, 146–152. [CrossRef]
3. Hashimoto, K. In pursuit of new corrosion resistant alloys. *Corrosion* **2002**, *58*, 715–722. [CrossRef]
4. Gostin, P.F. Corrosion Behaviour of Advanced Fe-Based Bulk Metallic Glasses. Ph.D. Thesis, Technische Universität Dresden, Dresden, Germany, May 2011. Available online: https://tud.qucosa.de/landing-page/?tx_dlf[id]=https%3A%2F%2Ftud.qucosa.de%2Fapi%2Fqucosa%253A25566%2Fmets (accessed on 20 December 2020).
5. Inoue, C.; Suryanarayana, A. Iron-based Bulk Metallic Glasses. *Int. Mater. Rev.* **2013**, *58*, 131–166.
6. Scully, J.R.; Gebert, A.; Payer, J.H. Corrosion and related mechanical properties of bulk metallic glasses. *J. Mater. Res.* **2007**, *22*, 302–313. [CrossRef]
7. Long, Z.L.; Shao, Y.; Deng, X.H.; Zhang, Z.C.; Jiang, Y.; Zhang, P.; Shen, B.L.; Inoue, A.A. Cr effects on magnetic and corrosion properties of Fe–Co–Si–B–Nb–Cr bulk glassy alloys with high glass-forming ability. *Intermetallics* **2007**, *15*, 1453–1458. [CrossRef]
8. Gostin, F.; Siegel, U.; Mickel, C.; Baunack, S.; Gebert, A.; Schultz, A.L. Corrosion behavior of the bulk glassy $(Fe_{44.3}Cr_5Co_5Mo_{12.8}Mn_{11.2}C_{15.8}B_{5.9})_{98.5}Y_{1.5}$ alloy. *J. Mater. Res.* **2009**, *24*, 1471–1479. [CrossRef]
9. Zhang, C.; Liu, L.; Chan, K.; Chen, Q.; Tang, C.Y. Wear behavior of HVOF-sprayed Fe-based Amorphous Coatings. *Intermetallics* **2012**, *29*, 80–85. [CrossRef]
10. Kiminami, C.S.; Souza, C.A.C.; Bonavina, L.F.; de Andrade Lima, L.R.P.; Suriñach, S.; Baró, M.D.; Bolfarini, C.; Botta, W.J. Partial crystallization and corrosion resistance of amorphous Fe-Cr-M-B (M = Mo, Nb) alloys. *J. Non-Cryst. Solids* **2010**, *356*, 2651–2657. [CrossRef]
11. Hashimoto, K. *Chemical Properties*; Butterworths: London, UK, 1983.
12. Waseda, Y.; Aust, A.K.T. Corrosion behavior of metallic glasses. *J. Mater. Sci.* **1981**, *16*, 2337–2359. [CrossRef]
13. Jayaraj, J.; Kim, Y.; Kim, K.; Seok, H.; Fleury, E. Corrosion behaviors of $Fe_{45-x}Cr_{18}Mo_{14}C_{15}B_6Y_2M_x$ (M = Al, Co, Ni, N and x = 0, 2) bulk metallic glasses under conditions simulating fuel cell environment. *J. Alloy. Compd.* **2007**, *434–435*, 237–239. [CrossRef]
14. Suryanarayana, C.; Inoue, A. *Bulk Metallic Glasses*; CRC Press: Boca Raton, FL, USA, 2011.
15. Jayaraj, J.; Kim, K.; Ahn, H.; Fleury, A.E. Corrosion mechanism of N-containing Fe–Cr–Mo–Y–C–B bulk amorphous alloys in highly concentrated HCl solution. *Mater. Sci. Eng. A* **2007**, *449–451*, 517–520. [CrossRef]
16. Jargelius-Pettersson, R. Electrochemical investigation of the influence of nitrogen alloying on pitting corrosion of austenitic stainless steels. *Corros. Sci.* **1999**, *41*, 1639–1664. [CrossRef]
17. Zhou, Z.; Wang, L.; Wang, F.; Zhang, H.; Liu, Y.; Xu, S. Formation and corrosion behavior of Fe-based amorphous metallic coatings by HVOF thermal spraying. *Surf. Coat. Technol.* **2009**, *204*, 563–570. [CrossRef]

18. Guo, S.; Chan, K.; Xie, S.; Yu, P.; Huang, Y.; Zhang, H. Novel centimeter-sized Fe-based bulk metallic glass with high corrosion resistance in simulated acid rain and seawater. *J. Non-Cryst. Solids* **2013**, *369*, 29–33. [CrossRef]
19. Zohdi, H.; Shahverdi, H.; Hadavi, S. Effect of Nb addition on corrosion behavior of Fe-based metallic glasses in Ringer's solution for biomedical applications. *Electrochem. Commun.* **2011**, *13*, 840–843. [CrossRef]
20. Tsai, P.; Xiao, A.; Li, J.; Jang, J.; Chu, J.; Huang, J. Prominent Fe-based bulk amorphous steel alloy with large supercooled liquid region and superior corrosion resistance. *J. Alloy. Compd.* **2014**, *586*, 94–98. [CrossRef]
21. Souza, C.; Ribeiro, D.; Kiminami, C. Corrosion resistance of Fe-Cr-based amorphous alloys: An overview. *J. Non-Cryst. Solids* **2016**, *442*, 56–66. [CrossRef]
22. Wang, S.L.; Yi, S. The corrosion behaviors of Fe-based bulk metallic glasses in a sulfuric solution at 70 degrees C. *Intermetallics* **2010**, *18*, 1950–1953. [CrossRef]
23. Asami, K.; Naka, M.; Hashimoto, T.; Masumoto, T. Effect of molybdenum on the anodic behavior of amorphous Fe–Cr–Mo–B alloys in hydrochloric acid. *J. Electrochem. Soc.* **1980**, *127*, 2130–2138. [CrossRef]
24. Marcus, P. On some fundamental factors in the effect of alloying elements on passivation of alloys. *Corros. Sci.* **1994**, *36*, 2155–2158. [CrossRef]
25. Pang, S.J.; Zhang, T.; Asami, K.; Inoue, A. Formation of bulk glassy $Fe_{75-x-y}Cr_xMo_yC_{15}B_{10}$ alloys and their corrosion behavior. *J. Mater. Res.* **2002**, *17*, 701–704. [CrossRef]
26. Habazaki, H.; Kawashima, A.; Asami, K.; Hashimoto, K. The Effect of Tungsten on the Corrosion Behavior of Amorphous Fe-Cr-W-P-C Alloys in 1M HCl. *J. Electrochem. Soc.* **1991**, *138*, 76–81. [CrossRef]
27. Habazaki, H.; Kawashima, A.; Asami, K.; Hashimot, K. The corrosion behavior of amorphous Fe-Cr-Mo-P-C and Fe-Cr-W-P-C alloys in 6 M HCl solution. *Corros. Sci.* **1992**, *33*, 225–236. [CrossRef]
28. Ma, X.H.; Zhang, L.; Yang, X.H.; Li, Q.; Huang, Y.D. Effect of Ni addition on corrosion resistance of FePC bulk glassy alloy. *Corros. Eng. Sci. Technol.* **2015**, *50*, 433–437. [CrossRef]
29. Souza, C. Corrosion resistance of amorphous and nanocrystalline Fe–M–B (M = Zr, Nb) alloys. *J. Non-Cryst. Solids* **2000**, *273*, 282–288. [CrossRef]
30. Pang, S.; Zhang, T.; Asami, K.; Inoue, A. Bulk glassy Fe–Cr–Mo–C–B alloys with high corrosion resistance. *Corros. Sci.* **2002**, *44*, 1847–1856. [CrossRef]
31. Wang, Z.M.; Ma, Y.T.; Zhang, J.; Hou, W.L.; Chang, X.C.; Wang, J.Q. Influence of yttrium as a minority alloying element on the corrosion behavior in Fe-based bulk metallic glasses. *Electrochim. Acta* **2008**, *54*, 261–269. [CrossRef]
32. Angelini, E.; Antonione, C.; Baricco, M.; Bianco, P.; Rosalbino, F.; Zucchi, F. Corrosion behaviour of $Fe_{80-x}Co_xB_{10}Si_{10}$ metallic glasses in sulphate and chloride media. *Werkst. Korros. Mater. Corros.* **1993**, *44*, 98–106. [CrossRef]
33. Li, Y.; Jia, X.; Zhang, W.; Fang, C.; Wang, X.; Qin, F.; Yamaura, S.; Yokoyama, a.Y. Effects of Alloying Elements on the Thermal Stability and Corrosion Resistance of an Fe-based

Metallic Glass with Low Glass Transition Temperature. *Metall. Mater. Trans. A* **2014**, *45A*, 2393–2398. [CrossRef]

34. Fang, H.; Chen, X.H.a.G. Effects of Mn addition on the magnetic property and corrosion resistance of bulk amorphous steels. *J. Alloy. Compd.* **2008**, *464*, 292–295. [CrossRef]

35. Masumoto, T.; Hashimoto, K.; Naka, M. Corrosion of amorphous alloys metals. In Proceedings of the 3rd International Conference on Rapidly Quenched Metals, London, UK, 3–7 July 1978.

36. Duarte, M.; Kostka, A.; Jimenez, J.; Choi, P.; Klemm, J.; Crespo, D.; Raabe, D.; Renner, F.U. Crystallization, phase evolution and corrosion of Fe-basedmetallic glasses: An atomic-scale structural and chemical characterization study. *Acta Mater.* **2014**, *71*, 20–30. [CrossRef]

37. Pang, S.J.; Zhang, T.; Asami, K.; Inoue, A. New Fe–Cr–Mo–(Nb, Ta)–C–B glassy alloys with high glass-forming ability and good corrosion resistance. *Mater. Trans.* **2001**, *42*, 376–379. [CrossRef]

38. Demaree, J.D.; Was, G.S.; Sorensen, N.R. Chemical and Structural Effects ofPhosphorus on the Corrosion Behavior of Ion Beam Mixed Fe-Cr-P Alloys. *J. Electrochem. Soc.* **1993**, *140*, 331–343. [CrossRef]

39. Naka, M.; Hashimoto, K.; Masumoto, T. Effect of Metalloid Elements on Corrosion Resistance of Amorphous Iron-Chromium Alloys. *J. Non. Cryst. Solids* **1978**, *28*, 403–413. [CrossRef]

40. Virtanen, S.; Bohni, B.E.a.H. Effect of Metalloid on the Passivity of Amorphous Fe-Cr Alloys. *J. Less-Common Metals* **1988**, *145*, 581–593. [CrossRef]

41. Chattoraj, I.; Baunack, S.; Stoica, M.; Gebert, A. Electrochemical response of $Fe_{65.5}Cr_4Mo_4Ga_4P_{12}C_5B_{5.5}$ bulk amorphous alloy in different aqueous media. *Mater. Corros.* **2004**, *55*, 36–42. [CrossRef]

42. Elsener, B.; Rossi, A. XPS Investigation of Passive Films on Amorphous Fe-Cr Alloys. *Electrochem. Acta* **1992**, *31*, 2269–2276. [CrossRef]

43. Pardo, A.; Otero, E.; Merino, M.C.; López, M.D.; Vázquez, M.; Agudo, P. The influence of Cr addition on the corrosion resistance of $Fe_{73.5}Si_{13.5}B_9Nb_3Cu_1$ metallic glass in SO_2 contaminated environments. *Corros. Sci.* **2001**, *43*, 689–705. [CrossRef]

44. Ha, H.M. Micro- and Nano-Scale Corrosion in Iron-Based Bulk Metallic Glass SAM 1651 and Silvered-Cored MP35N LT Composite. Ph.D. Thesis, Case Western Reserve University, Cleveland, OH, USA, 2010.

45. Li, H.; Yi, S. Corrosion behaviors of bulk metallic glasses $Fe_{66.7}C_{7.0}Si_{3.3}B_{5.5}P_{8.7}Cr_{2.3}Al_{2.0}Mo_{4.5}$ having different crystal volume fractions. *Mater. Chem. Phys.* **2008**, *112*, 305–309. [CrossRef]

46. Naka, M.; Masumoto, K.H.a.T. Effect of Heat Treatment on Corrosion Behavior of Amorphous Fe-Cr-P-C and Fe-Ni-Cr-P-B Alloys in 1N HCl. *Corrosion* **1980**, *36*, 679–686. [CrossRef]

47. Gostin, P.; Gebert, A.; Schultz, L. Comparison of the corrosion of bulk amorphous steel with conventional steel. *Corros. Sci.* **2010**, *52*, 273–281. [CrossRef]

48. Pardo, A.; Merino, M.C.; Otero, E.; López, M.D.; M'hich, A. Influence of Cr additions on corrosion resistance of Fe- and Co-based metallic glasses and nanocrystals in H_2SO_4. *J. Non-Cryst. Solids* **2006**, *352*, 3179–3190. [CrossRef]
49. Yang, Y.; Zhang, C.; Peng, Y.; Yu, Y.; Liu, L. Effects of crystallization on the corrosion resistance of Fe-based amorphous coatings. *Corros. Sci.* **2012**, *59*, 10–19. [CrossRef]
50. Lin, L.; Zhang, C. Fe-based amorphous coatings: Structures and properties. *Thin Solid Film.* **2014**, *561*, 70–86.
51. Ha, H.; Miller, J.R.; Payer, J. Devitrification of Fe-Based Amorphous Metal SAM 1651 and the Effect of Heat-Treatment on Corrosion Behavior. *ECS Electrochem. Soc.* **2009**, *156*, C246–C252. [CrossRef]
52. Liu, D.; Zhang, H.F.; Hu, Z.Q.; Gaoa, W. Magnetic and corrosion properties of $Fe_{56}Co_7M_2Mo_5Zr_{10}B_{20}$ (M =W or Ni) bulk metallic glasses. *J. Alloy. Compd.* **2006**, *422*, 28–31. [CrossRef]
53. Zhang, L.; Ma, X.; Li, Q.; Zhang, J.; Dong, Y.; Chang, C. Preparation and properties of $Fe_{80-x}Ni_xP_{14}B_6$ bulk metallic glasses. *J. Alloy. Compd.* **2014**, *608*, 79–84. [CrossRef]
54. Hashimoto, T.M.a.K. Corrosion Properties of Amorphous Metals. *J. Phys. Colloq.* **1980**, *41*, 894–900.
55. Wang, L.; Chao, Y. Corrosion behavior of $Fe_{41}Co_7Cr_{15}Mo_{14}C_{15}B_6Y_2$ bulk metallic glass in NaCl solution. *Mater. Lett.* **2012**, *69*, 76–78. [CrossRef]
56. Guo, R.; Zhang, C.; Chen, Q.; Yang, Y.; Li, N.; Liu, L. Study of structure and corrosion resistance of Fe-based amorphous coatings prepared by HVAF and HVOF. *Corros. Sci.* **2011**, *53*, 2351–2356. [CrossRef]
57. Shi, M.; Pang, S.; Zhang, T. Towards improved integrated properties in FeCrPCB bulk metallic glasses by Cr addition. *Intermetallics* **2015**, *61*, 16–20. [CrossRef]
58. Wang, S.; Li, H.; Zhang, X.; Yi, S. Effects of Cr contents in Fe-based bulk metallic glasses on the glass forming ability and the corrosion resistance. *Mater. Chem. Phys.* **2009**, *878*, 113. [CrossRef]
59. Pang, S.; Zhang, T.; Asami, K.; Inoue, A. Synthesis of Fe–Cr–Mo–C–B–P bulk metallic glasses with high corrosion resistance. *Acta Mater.* **2002**, *50*, 489–497. [CrossRef]
60. Emran, K.M. Effects of concentration and temperature on the corrosion properties of the Fe-Ni-Mn alloy in HCl solutions. *Res. Chem. Intermed.* **2015**, *41*, 3583–3596. [CrossRef]
61. Chang, C.; Qin, C.; Makino, A.; Inoue, A. Enhancement of glass-forming ability of FeSiBP bulk glassy alloys with good soft-magnetic properties and high corrosion resistance. *J. Alloys Compound.* **2012**, *533*, 67–70. [CrossRef]
62. Mariano, N.A.; Souza, C.A.C.; May, J.E.; Kuri, S.E. Influence of Nb content on the corrosion resistance and saturation magnetic density of FeCuNbSiB alloys. *Mat. Sci. Eng. A* **2003**, *A354*, 1–5. [CrossRef]
63. Fan, H.; Zheng, W.; Wang, G.; Liaw, P.; Shen, A.J. Corrosion Behavior of $Fe_{41}Co_7Cr_{15}Mo_{14}C_{15}B_6Y_2$ Bulk Metallic Glass in Sulfuric Acid Solutions. *Metall. Mater. Trans. A* **2011**, *42A*, 1524–1533. [CrossRef]
64. Rebak, R.B.; Day, S.D.; Lian, T.; Farmer, J.C. Environmental Testing of Iron-Based Amorphous Alloys. *Metall. Mater. Trans. A* **2008**, *39A*, 225–234. [CrossRef]

65. Lekatou, A.; Marinou, A.; Patsalas, P.; Karakassides, M. Aqueous corrosion behaviour of Fe–Ni–B metal glasses. *J. Alloy. Compd.* **2009**, *483*, 514. [CrossRef]
66. Long, Z.; Shao, Y.; Xie, G.; Zhang, P.; Shen, B.; Inoue, A. Enhanced soft-magnetic and corrosion properties of Fe-based bulk glassy alloys with improved plasticity through the addition of Cr. *J. Alloy. Compd.* **2008**, *462*, 52. [CrossRef]
67. Wang, Y.; Li, H.; Cheng, Y.; Wei, S.; Zheng, Y. Corrosion performances of a Nickel-free Fe-based bulk metallic glass in simulated body fluids. *Electrochem. Commun.* **2009**, *11*, 2187–2190. [CrossRef]
68. Wang, Y.; Li, H.; Zheng, Y.; Li, M. Corrosion performances in simulated body fluids and cytotoxicity evaluation of Fe-based bulk metallic glasses. *Mater. Sci. Eng. C* **2012**, *32*, 599–606. [CrossRef]
69. Calin, M.; Gebert, A.; Ghinea, A.; Gostin, P.; Abdi, S.; Mickel, C.; Eckert, J. Designing biocompatible Ti-based metallic glasses for implant applications. *Mater. Sci. Eng. C* **2013**, *33*, 875–883. [CrossRef]
70. Li, S.; Wei, Q.; Li, Q.; Jiang, B.; Chen, Y.; Sun, Y. Development of Fe-based bulk metallic glasses as potential biomaterials. *Mater. Sci. Eng. C* **2015**, *52*, 235–241. [CrossRef]
71. Ma, X.; Zhen, N.; Guo, J.; Li, Q.; Chang, C.; Sun, Y. Preparation of Ni-based bulk metallic glasses with high corrosion resistance. *J. Non-Cryst. Solids* **2016**, *443*, 91–96. [CrossRef]
72. Tan, C.; Jiang, W.; Zhang, Z.; Wu, X.; Lin, J. The effect of Ti-addition on the corrosion behavior of the partially crystallized Ni-based bulk metallic glasses. *Mater. Chem. Phys.* **2008**, *108*, 29–32. [CrossRef]
73. Yamaura, S.; Yokoyama, M.; Inoue, A. Development of the Ni-based Metallic Glassy Bipolar Plates for Proton Exchange Membrane Fuel Cell (PEMFC). *J. Phys. Conf. Ser.* **2009**, *144*, 1–4. [CrossRef]
74. Hashimoto, K.; Shinomiya, H.; Nakazawa, A.; Kato, Z.; El-Moneim, A.A.; Niizeki, Y.; Asami, K. Corrosion-resistance Bulk Amorphous Ni-Cr-Ta-Mo-Nb-P Alloys in Concentarted Hydrochloric Acids. *ECS Trans.* **2006**, *1*, 49–59. [CrossRef]
75. Zhu, Z.; Zhang, H.; Ding, B.; Hu, Z. Synthesis and properties of bulk metallic glasses in the ternary Ni–Nb–Zr alloy system. *Mater. Sci. Eng. A* **2008**, *492*, 221–229. [CrossRef]
76. Tien, H.Y.; Lin, C.Y.; Chin, T.S. New ternary Ni–Ta–Sn bulk metallic glasses. *Intermetallics* **2006**, *14*, 1075–1078. [CrossRef]
77. Wang, Z.M.; Zhang, J.; Chang, X.C.; Wang, W.L.H.a.J.Q. Structure inhibited pit initiation in a Ni–Nb metallic glass. *Corros. Sci.* **2010**, *52*, 1342–1350. [CrossRef]
78. Wang, A.P.; Chang, X.; Hou, W.; Wang, J. Corrosion behavior of Ni-based amorphous alloys and their crystalline counterparts. *Corros. Sci.* **2007**, *49*, 2628–2635. [CrossRef]
79. Wang, P.; Zhang, T.; Wang, J.Q. Formation and Properties of Ni-Based Amorphous Metallic Coating Produced by HVAF Thermal Spraying. *Mater. Trans.* **2005**, *46*, 1010–1015. [CrossRef]
80. Wang, P.; Zhan, T.; Wang, J.Q. Ni-based fully amorphous metallic coating with high corrosion resistance. *Philos. Mag. Lett.* **2006**, *86*, 5–11. [CrossRef]

81. Zhu, C.; Wang, Q.; Wang, Y.; Qiang, J.; Dong, C. Ni-based B–Fe–Ni–Si–Ta bulk metallic glasses designed using cluster line, minor alloying, and element substitutions. *Intermetallics* **2010**, *18*, 791–795. [CrossRef]
82. Shinomiya, H.; Kato, Z.; Hashimoto, K. Effects of Corrosion-Resistant Elements on the Corrosion Resistance of Amorphous Bulk Ni-Cr-Mo-Ta-Nb-4P Alloys in Concentrated Hydrochloric Acids. *ECS Trans.* **2009**, *16*, 9–18. [CrossRef]
83. Katagiri, H.; Meguro, S.; Yamasaki, M.; Habazaki, H.; Sato, T.; Kawashima, A.; Hashimoto, K. Synergistic effect of three corrosion-resistant elements on corrosion resistance in concentrated hydrochloric acid. *Corros. Sci.* **2001**, *43*, 171–182. [CrossRef]
84. Katagiri, H.; Meguro, S.; Yamasaki, M.; Habazaki, H.; Sato, T.; Kawashima, A.; Hashimoto, K. An attempt at preparation of corrosion-resistant bulk amorphous Ni-Cr-Ta-Mo-P-B alloys. *Corros. Sci.* **2001**, *43*, 183–191. [CrossRef]
85. Qin, C.; Zeng, Y.; Louzguine, D.; Nishiyama, N.; Inoue, A. Corrosion resistance and XPS studies of Ni-rich Ni–Pd–P–B bulk glassy alloys. *J. Alloys Compound.* **2010**, *504S*, 172–175. [CrossRef]
86. Pang, S.; Zhang, T.; Asami, K.; Inoue, A. Bulk glassy Ni(Co–)Nb–Ti–Zr alloys with high corrosion resistance and high strength. *Mater. Sci. Eng. A* **2004**, *368*, 375–377. [CrossRef]
87. Zhang, T.; Pang, S.; Asami, K.; Inoue, A. Glassy Ni-Ta-Ti-Zr(-Co) Alloys with High Thermal Stability. *Mater. Trans.* **2003**, *44*, 2322–2325. [CrossRef]
88. Kawashima, A.; Sato, T.; Asami, K. Characterization of Surface of Amorphous Ni-Nb-Ta-P Alloys Passivated in a 12 kmol/m^3 HCl Solution. *Mater. Trans.* **2004**, *45*, 131–136. [CrossRef]
89. Qin, C.L.; Asami, K.; Kimura, H.; Zhang, W.; Inoue, A. High Corrosion Resistant Ni-Based Glassy Alloys in Boiling Nitric Acid Solutions. *Mater. Trans.* **2009**, *50*, 1304–1307. [CrossRef]
90. Kawashima, A.; Hashimoto, K. Highly corrosion-resistant Ni-based bulk amorphous alloys. *Mater. Sci. Eng. A* **2001**, *304–306*, 753–757. [CrossRef]
91. Pang, S.; Shek, C.; Asami, K. Formation and corrosion behavior of glassy Ni–Nb–Ti–Zr–Co(–Cu) alloys. *J. Alloys Compound.* **2007**, *434–435*, 240–243. [CrossRef]
92. Lee, H.J.; Akiyama, E.; Habazaki, H.; Kawashima, A.; Hashimoto, K. The corrosion Behavior of Ni-Ta-5P Alloys in Concentrated Hydrochloric Acid. *Mater. Trans. JIM* **1996**, *37*, 383–388. [CrossRef]
93. Lee, H.-J.; Akiyama, E.; Habazaki, H.; Kawashima, A.; Hashimoto, K. The Roles of Tantalum and Phosphorus in the Corrosion Behavior of Ni-Ta-P Alloya in 12M HCl. *Corros. Sci.* **1997**, *39*, 321–332. [CrossRef]
94. Lee, H.J.; Akiyama, E.; Habazaki, H.; Kawashima, A.; Hashimoto, K. The Corrosion Behavior of Amorphous and Crystalline Ni-10Ta-20P Alloys in 12 M HCl. *Corros. Sci.* **1996**, *38*, 1269–1279. [CrossRef]
95. Raicheff, R.; Zaprianova, V. Effect of crystallization on the electrochemical corrosion behavior of some nickel-based amorphous alloys. *J. Mater. Sci. Lettt.* **2000**, *19*, 3–5. [CrossRef]

96. Liu, S.; Huang, L.; Zhang, T. Effects of crystallization on corrosion behaviors of a Ni-based bulk metallic glass. *Int. J. Miner. Metall. Mater.* **2012**, *19*, 146–150. [CrossRef]
97. Zhang, K.; Gao, X.; Dong, Y.; Xing, Q.; Wang, Y. Effect of annealing on the microstructure, microhardness, and corrosion resistance of Ni62Nb33Zr5 metallic glass and its composites. *J. Non-Cryst. Solids* **2015**, *425*, 46–51. [CrossRef]
98. Tam, M.; Shek, C. Crystallization and corrosion resistance of $Cu_{50}Zr_{45}Al_5$ bulk amorphous alloy. *Mater. Chem. Phys.* **2006**, *100*, 34–37. [CrossRef]
99. Souza, C.; Politi, F.; Kiminami, C. Influence of structural relaxation and partial devitrification on the corrosion resistance of $Fe_{78}B_{13}Si_9$ amorphous alloy. *Scr. Mater.* **1998**, *39*, 329–334. [CrossRef]
100. Habazaki, H.; Ukai, H.; Hashimoto, K. Corrosion behavior of amorphous Ni—Cr—Nb–P–B bulk alloys in 6M HCl solution. *Mater. Sci. Eng. A* **2001**, *318*, 77–86. [CrossRef]
101. Wang, P.; Chang, X.; Wang, J. Preparation and corrosion behavior of amorphous Ni-based alloy coatings. *Mater. Sci. Eng. A* **2007**, *449–451*, 277–280. [CrossRef]
102. Asami, K.; Habazaki, H.; Hashimoto, K. Recent Development of Highly Corrosion Resistant Bulk Glassy Alloys. *Mater. Sci. Forum* **2005**, *502*, 225–230. [CrossRef]
103. Zeng, Y.; Qin, C.; Inoue, A. New nickel-based bulk metallic glasses with extremely high nickel content. *J. Alloy. Compd.* **2010**, *489*, 80–83. [CrossRef]
104. Vasantha, V.; Fleury, E. Corrosion properties of Ni-Nb & Ni-Nb-M (M = Zr, Mo, Ta & Pd) metallic glasses in simulated PEMFC conditions. *J. Phys. Conf. Ser.* **2009**, *144*, 1–4.
105. Al-Refai, H. Resistivity and Passivity Characterization of Ni-Base Glassy Alloys in NaOH Media. *Metals* **2018**, *8*, 1–13.
106. Amiya, I.K. Fe-(Cr,Mo)-(C,B)-Tm Bulk Metallic Glasses with High Strength and High Glass-Forming Ability. *Mater. Trans.* **2006**, *47*, 1615–1618. [CrossRef]
107. Inoue, A.; Shen, B.; Koshiba, H.; Kato, H.; Yavari, A. Ultra-high strength above 5000 MPa and soft magnetic properties of Co–Fe–Ta–B bulk glassy alloys. *Acta Mater.* **2004**, *52*, 1631–1637. [CrossRef]
108. Shen, B.; Pang, S.; Zhang, T.; Kimura, H. Corrosion properties of $Co_{43}Fe_{20}Ta_{5.5}B_{31.5}$ bulk glassy alloy. *J. Alloy. Comp.* **2008**, *460*, 11–13. [CrossRef]
109. Zhu, C.; Wang, Q.; Wang, Y.; Qiang, J.; Dong, C. Co–B–Si–Ta bulk metallic glasses designed using cluster line and alloying. *J. Alloy. Compd.* **2010**, *504*, 34–37. [CrossRef]
110. Pardo, A.; Otero, E.; Merino, M.C.; López, M.D.; Vázquez, M.; M'hich, A. Influence of Chromium Addition on the Corrosion Resistance $Co_{73.5}Si_{13.5}B_9Nb_3Cu_1$ Metallic Glass in Sodium Sulfate. *Corrosion* **2002**, *58*, 987–994. [CrossRef]
111. Lotfollahi, Z.; García-Arribas, A.; Amirabadizadeh, A.; Kurlyandskaya, G. Comparative study of magnetic and magnetoim pedance properties of CoFeSiB-based amorphous ribbons of the same geometry with Mo or W additions. *J. Alloy. Compd.* **2017**, *693*, 767–776. [CrossRef]
112. Zhoua, Z.; Wei, Q.; Li, Q.; Jiang, B.; Chen, Y.; Sun, Y. Development of Co-based bulk metallic glasses as potential biomaterials. *Mater. Sci Eng. C* **2016**, *69*, 46–51. [CrossRef]

113. Marzo, F.; Pierna, A.; Barranco, J.; Vara, G.; Lorenzo, A.; García, J. Corrosion Behaviour of Fe/Co Based Amorphous Metallic Alloys in Saline Solutions: New Materials for GMI Based Biosensors. *Port. Electrochim. Acta* **2007**, *25*, 131–137. [CrossRef]
114. Nowosielski, R.; Zajdel, A.; Lesz, S. Influence of crystallisation anamorphous $Co_{77}Si_{11.5}B_{11.5}$ alloy on corrosion behavior. *J. Of Achiev. Int. Mater. Manuf. Eng.* **2007**, *20*, 167–170.
115. Gu, Y.; Zheng, Z.; Niu, S.; Ge, W.; Wang, Y. The seawater corrosion resistance and mechanical properties of $Cu_{47.5}Zr_{47.5}Al_5$ bulk metallic glass and its composites. *J. Non-Cryst. Solids* **2013**, *380*, 135–140. [CrossRef]
116. Nie, X.; Yang, X.; Jiang, J. Ti microalloying effect on corrosion resistance and thermal stability of CuZr-based bulk metallic glasses. *J. Alloy. Compd.* **2009**, *481*, 498–502. [CrossRef]
117. Qin, C.; Zhang, W.; Asami, K.; Kimura, H.; Wang, X.; Inoue, A. A novel Cu-based BMG composite with high corrosion resistance and excellent mechanical properties. *Acta Mater.* **2006**, *54*, 3713–3719. [CrossRef]
118. Tam, M.; Pang, S.; Shek, C. Corrosion behavior and glass-forming ability of Cu–Zr–Al–Nb alloys. *J. Non-Cryst. Solid* **2007**, *353*, 3596–3599. [CrossRef]
119. Qin, C.; Zhang, W.; Kimura, H.; Asami, K.; Inoue, A. New Cu-Zr-Al-Nb Bulk Glassy Alloys with High Corrosion Resistance. *Mater. Trans.* **2004**, *45*, 1958–1961. [CrossRef]
120. Qin, C.; Asami, K.; Zhang, T.; Zhang, W.; Inoue, A. Corrosion Behavior of Cu-Zr-Ti-Nb Bulk Glassy Alloys. *Mater. Trans.* **2003**, *44*, 749–753. [CrossRef]
121. Qin, C.; Zhang, W.; Zhang, Q.; Asami, K.; Inoue, A. Electrochemical properties and surface analysis of Cu–Zr–Ag–Al–Nb bulk metallic glasses. *J. Alloy. Compd.* **2009**, *483*, 317–320. [CrossRef]
122. Asami, K.; Qin, C.-L.; Inoue, A. Effect of additional elements on the corrosion behavior of a Cu–Zr–Ti bulk metallic glass. *Mater. Sci. Eng. A* **2004**, *375–377*, 235–239. [CrossRef]
123. Qin, C.; Asami, K.; Zhang, T.; Inoue, A. Effects of Additional Elements on the Glass Formation and Corrosion Behavior of Bulk Glassy Cu–Hf–Ti Alloys. *Mater. Trans.* **2003**, *44*, 1042–1045. [CrossRef]
124. Yamamoto, T.; Qin, C.; Zhang, T.; Inoue, A. Formation, Thermal Stability, Mechanical Properties and Corrosion Resistance of Cu–Zr–Ti–Ni–Nb Bulk Glassy Alloys. *Mater. Trans.* **2003**, *44*, 1147–1152. [CrossRef]
125. Qin, C.; Inoue, A. Glass Formation, Chemical Properties and Surface Analysis of Cu-Based Bulk Metallic Glasses. *Int. J. Mol. Sci.* **2011**, *12*, 2275–2293. [CrossRef]
126. Qin, C.L.; Zhang, W.; Asami, K.; Inoue, A. Influence of alloying elements Ni and Nb on thermal stability and corrosion resistance of Cu-based bulk metallic glasses. *J. Mater. Res.* **2007**, *22*, 1710–1717. [CrossRef]
127. Qin, C.; Wang, L.; Wang, Z.; Ding, J.; Inoue, A. Synthesis, properties and XPS analysis of Cu-, Ni-, Ti-based metallic glasses. *Spec. Rep.* **2014**, *11*, 247–257.
128. Lin, H.; Wu, J.; Wang, C.; Lee, P. The corrosion behavior of mechanically alloyed Cu–Zr–Ti bulk metallic glasses. *Mater. Lett.* **2008**, *62*, 2995–2998. [CrossRef]

129. Liu, L. Improvement of corrosion resistance of Cu-based bulk metallic glasses by the microalloying of Mo. *Intermetallics* **2007**, *15*, 679–682. [CrossRef]
130. Liu, L. The effect of microalloying on thermal stability and corrosion resistance of Cu-based bulk metallic glasses. *Mater. Sci. Eng. A* **2006**, *415*, 286–290. [CrossRef]
131. Liu, L.; Liu, B. Influence of the micro-addition of Mo on glass forming ability and corrosion resistance of Cu-based bulk metallic glasses. *Electrochim. Acta* **2006**, *51*, 3724–3730. [CrossRef]
132. Chen, S.; Lin, S.; Chen, J.; Lin, Y. Thermal stability and corrosion behavior of Cu-Zr-Al-Y bulk metallic glass. *Intermetallics* **2010**, *18*, 1954–1957. [CrossRef]
133. Jinhong, P.; Ye, P.; Xiancong, H. Influence of Minor Addition of In on Corrosion Resistance of Cu-Based Bulk Metallic Glasses in 3.5% NaCl Solution. *Rare Met. Mater. Eng.* **2014**, *43*, 32–35. [CrossRef]
134. Zhang, C.; Qiu, N.; Kong, L.; Li, H. Thermodynamic and structural basis for electrochemical response of Cu–Zr based metallic glass. *J. Alloy Compd.* **2015**, *645*, 487–490. [CrossRef]
135. Zhang, C.; Wang, J.; Qiu, N.; Li, H. Cerium addition on pitting corrosion of $(Cu_{50}Zr_{50})_{100-2x}Ce_{2x}$ (x=0, 1, 2 and 3) metallic glasses in seawater. *J. Rare Earths* **2015**, *33*, 102–106. [CrossRef]
136. Lu, H.B.; Zhang, L.C.; Schultz, L. Pitting corrosion of Cu–Zr metallic glasses in hydrochloric acid solutions. *J. Alloy. Compd.* **2008**, *462*, 60–67. [CrossRef]
137. Zander, D.; Gallino, I. Corrosion resistance of Cu–Zr–Al–Y and Zr–Cu–Ni–Al–Nb bulk metallic glasses. *J. Alloy. Compd.* **2007**, *434–435*, 234–236. [CrossRef]
138. An, W.K.; Cai, A.h.; Xiong, X.; Liu, Y.; Luo, Y.; Li, T.l.L.a.X.S. Corrosion Behavior of $Cu_{60}Zr_{30}Ti_{10}$ Metallic Glass in the Cl− Containing Solution. *Mater. Sci. Appl.* **2011**, *2*, 546–554.
139. Cai, H.; Xiong, X.; Liu, Y.; An, W.K.; Zhou, G.J.; Li, Y.L.a.T. Corrosion behavior of $Cu_{55}Zr_{35}Ti_{10}$ metallic glass in the chloride media. *Mater. Chem. Phys.* **2012**, *134*, 938–944. [CrossRef]
140. Vincent, S.; Khan, A.F.; Bhatt, J. Corrosion characterization on melt spun $Cu_{60}Zr_{20}Ti_{20}$ metallic glass: An experimental case study. *J. Non-Cryst. Solids* **2013**, *379*, 48–53. [CrossRef]
141. Xu, T.; Pang, S.; Zhang, T. Glass formation, corrosion behavior, and mechanical properties of novel Cr-rich Cr–Fe–Mo–C–B–Y bulk metallic glasses. *J. Alloy. Compd.* **2015**, *625*, 318–322. [CrossRef]
142. Si, J.; Chen, X.; Cai, Y.; Wu, Y.; Wang, T.; Hui, X. Corrosion behavior of Cr-based bulk metallic glasses in hydrochloric acid solutions. *Corros. Sci.* **2016**, *107*, 123–132. [CrossRef]

5. Low-Density Metallic Glassess

Low-density metallic glasses based on Ti, Mg, Al, and Ca are attractive for wide ranging structural applications because of their low density and high specific strength and may be particularly suitable for bioimplant applications because of their superior mechanical properties and biocompatibility [1,2]. Understanding the corrosion behavior of low-density metallic glasses will help in the development of next generation alloys where weight constraints and surface degradation resistance are critical to meet the design goals.

5.1. Titanium (Ti)-Based Metallic Glasses

The electrochemical behavior of different grades of Ti-based bulk metallic glasses (BMGs) has been widely studied in acidic, alkaline, and simulated body fluid environments [3–6]. Surface wettability is an important consideration for orthopedic and dental implants because it controls cell response and affects the cytocompatibility of implanted biomaterials [2]. Ti-based BMGs have been reported to be bioactive and promote bone growth [7]. In this regard, apatite formation is important in simulating nucleation and growth of cells on bio-active implants due to its mineral and chemical similarity to natural bone. The interaction between the biomedical implant and the tissue includes both the host response as well as implant material response [8]. Other considerations include the release of toxic elements and the cost of production. Alloying elements such as nickel (Ni), beryllium (Be), chromium (Cr), vanadium (V), and aluminum (Al) are known to improve the glass-forming ability of Ti-based BMGs. However, these elements are not desirable because of high toxicity [6,9,10]. So far, a wide range of Ti-based BMGs have been developed without these toxic elements including Ti-Zr-Fe-(Ta, Pd, Nb)-Si [4,9,11,12], Ti-Zr-Fe-Si-Mo-Nb [12], Ti-Zr-Cu-Pd [13–15], and Ti-Zr-Cu-Pd-Sn [10,16–18]. These Ti-based BMGs form a protective passive film with good corrosion resistance hindering metal-ion release and improving biocompatibility [8]. $Ti_{46}Cu_{27.5}Zr_{11.5}Co_7Sn_3Ag_4Si_1$ and $Ti_{44.1}Zr_{9.8}Pd_{9.8}Cu_{30.38}Sn_{3.92}Nb_2$ [10] BMGs showed higher corrosion potentials and lower corrosion current densities as compared with Ti-6Al-4V in electrolytes such as HCl, NaCl, NaOH, PBS, simulated body fluid, and Hank's balanced salt solution [6]. All of these alloys exhibited spontaneous passivation in the above mentioned environments with a lower passive current density than pure Ti and Ti-6Al-4V.

The mismatch of Young's modulus between human bone (10–30 GPa) and Ti-based BMGs (around 95 GPa) raises concern about stress-shielding. This issue may be addressed by synthesizing porous structures as demonstrated for Ti-based BMGs using spark plasma sintering process. Corrosion behavior of a porous Ti-based BMG was compared with the bulk non-porous counterpart and commercial Ti alloys [19]. Continuous increase in corrosion current density of porous $Ti_{45}Zr_{10}Cu_{31}Pd_{10}Sn_4$ BMG in the anodic part of the polarization curve was attributed to crevice corrosion [19]. In another study, $Ti_{45}Zr_{10}Cu_{31}Pd_{10}Sn_4$ BMG powder was synthesized by high-pressure argon gas atomization and subsequently, spark plasma sintered (SPS). The sintering temperature was found to influence the passivation and pitting potential of the alloy in Hank's solution while partial crystallization decreased the stability of the passive film [20].

5.1.1. Effect of Alloying Elements

Chemical composition plays an important role in determining the corrosion resistance and biocompatibility of Ti-based BMGs. Release of Be, Al, V, Cr, Ni, and Cu may lead to toxicity in the human body [9]. Release of Cu from $Zr_{50}Cu_{43}Al_7$ BMG after 1-day immersion in a cell culture medium (DMEM) was below 50 ppb [21], indicating that Cu addition in certain Ti-based BMGs may not be a serious concern. Increasing Ti content in Ti–Zr–Si thin film metallic glasses led to the formation of a thick and dense protective TiO_2 layer and high corrosion resistance in SBF [22]. Electrochemical studies for Ti-Zr-Cu-Fe-Sn-Si amorphous alloys in PBS revealed that increasing the (Ti+Zr)/Cu ratio increased the pitting potential and passive film stability due to the formation of a chemically stable and structurally dense protective film of TiO_2 and ZrO_2 [2]. The presence of Niobium (Nb), Hafnium (Hf), Tin (Sn), and Tantalum (Ta) improved the electrochemical properties of Ti-based BMGs in different environments due to the formation of a passive film on the surface [5,9,23–28]. The corrosion current density of Ti-based BMGs with Ta addition was an order of magnitude lower compared to the base alloy without Ta [23]. $Ti_{41.5}Zr_{2.5}Hf_5Cu_{37.5}Ni_{7.5}Si_1Sn_5$ BMG is one of the best Be-free bulk-glass formers with a large critical casting thickness. In addition to excellent biocompatibility, this amorphous alloy showed very good corrosion resistance in SBF due to a complex oxide surface film consisting of TiO_2, ZrO_2, and HfO_2 [28]. The addition of noble metals like gold (Au) and platinum (Pt) to Ti-based BMGs was found to be beneficial in terms of spontaneous passivity and enhancement in corrosion resistance [26]. Furthermore, Ti-Zr-Pd-Si glassy alloys with Pd addition exhibited better corrosion resistance than pure Ti and Ti-6Al-4V in lactic acid and PBS [4].

5.1.2. Effects of Structure and Crystallinity

Partial crystallization of $Ti_{40}Zr_{10}Cu_{36}Pd_{14}$ bulk metallic glass increased pitting potential and corrosion resistance as compared to as-cast and fully crystallized alloys due to the formation of stable and protective passive films. The formation of a nanocrystalline Ti_3Cu_4 phase after annealing of $Ti_{40}Zr_{10}Cu_{36}Pd_{14}$ BMG caused the enrichment of Pd in the matrix and promoted the formation of a protective passive film. As a result of the numerous nanocrystals, uniform passivity was maintained in the partially crystallized alloy. However, the fully crystallized alloy showed the lowest corrosion resistance due to the formation of other phases such as Ti_2Pd and Ti_2Pd_3 in addition to Ti_3Cu_4, which resulted in micro-galvanic corrosion between Cu-rich and Pd-rich phases [29,30]. In another study, nano-crystallized $Ti_{42}Zr_{40}Si_{15}Ta_3$ metallic glass with different degrees of α-Ti nanophases displayed enhanced pitting and corrosion resistance [31]. Galvanic corrosion did not occur in the partially crystallized Ti-based BMG with α-Ti nanocrystalline phase because α-Ti was inert and did not change the homogenous structure of the passive film.

5.2. Ti-Based Bulk Metallic Glass Composites

To improve the tensile ductility and toughness of bulk metallic glasses, in situ BMG composites have been developed consisting of crystalline dendrites in an amorphous matrix to arrest the propagation of shear bands and prevent catastrophic failure [32]. However, the presence of a crystalline second phase in the amorphous matrix has been reported to reduce the corrosion resistance of the alloys due to galvanic coupling and surface heterogeneities which act as initiation sites for localized corrosion [33]. To date, several Ti-based BMG composites have been developed, but there are limited reports on their corrosion behavior [3,32–35]. In a recent study, various amounts of Nb was added to $Ti_{45}Zr_{16}Be_{20}Cu_{10}Ni_9$ metallic glass alloy to stabilize a crystalline β-phase [33]. The composition variation directly affected the corrosion current density of the BMG composite in H_2SO_4 at 80 °C. Potentiodynamic studies showed that the best corrosion behavior was seen for the composite containing 10 at. % of Nb. A higher fraction of Nb (15 at. %) led to second phase inhomogeneity in the amorphous matrix, decreasing the corrosion resistance [33]. A reduction in corrosion resistance for alloys with higher Nb content was related to preferential dissolution of the matrix due to the galvanic effect between the matrix and the dendrite phase. Early transition metals (V, Ta, Nb) have been added to Ti-based BMGs to form quasi-crystalline phases and bulk metallic glass matrix composites (BMGMCs). The Ti-Zr-Be-Cu-Ni amorphous alloy with the addition of V/Ta/Nb exhibited better corrosion resistance in the potential range up to 1.5 V vs. SCE in 1 M H_2SO_4 + 2 ppm F^- at 80 °C while at higher potential ranges, Ti-6Al-4V showed better corrosion resistance. The alloy with 3 at. % Ta had the best corrosion resistance due to the formation of a strong and stable passive layer retarding the detrimental effect of inhomogeneity introduced by the presence of i-phase [32]. The corrosion parameters of these alloys obtained from potentiodynamic polarization tests in H_2SO_4 at 80 °C are summarized in Table 5.1 [32].

Table 5.1. Corrosion current density (i_{corr}), corrosion potential (E_{corr}), and polarization resistance (R_p) of Ti-Zr-based BMGs with the addition of V, Ta, and Nb, in comparison with a Ti-Al-V crystalline alloy, obtained from potentiodynamic polarization in 1 M H_2SO_4 + 2 ppm F^- with air bubbling [32].

Alloy Compositions	i_{corr} (μA/cm^2)	E_{corr} (V)	R_p (ohm)
$Ti_{40}Zr_{29}Be_{16}Cu_8Ni_7$	5.46 ± 1.03	−0.216 ± 0.0184	2254.0 ± 128.69
$(Ti_{40}Zr_{29}Be_{16}Cu_8Ni_7)_{99}V_1$	8.66 ± 1.80	−0.127 ± 0.0163	366.0 ± 113.14
$(Ti_{40}Zr_{29}Be_{16}Cu_8Ni_7)_{97}V_3$	8.11 ± 0.54	−0.249 ± 0.0141	1298.8 ± 619.78
$(Ti_{40}Zr_{29}Be_{16}Cu_8Ni_7)_{95}V_5$	7.04 ± 1.73	−0.240 ± 0.0324	1551.67 ± 565.34
$(Ti_{40}Zr_{29}Be_{16}Cu_8Ni_7)_{97}Ta_3$	4.26 ± 1.19	−0.284 ± 0.0007	1079.0 ± 213.55
$(Ti_{40}Zr_{29}Be_{16}Cu_8Ni_7)_{95}Nb_5$	6.31 ± 2.47	−0.223 ± 0.0396	3003.5 ± 419.31
$Ti_{86.20}Al_{10.20}V_{3.60}$	599.5 ± 236.88	−0.745 ± 0.0488	8.7 ± 4.38

The corrosion potential and current density of $Ti_{62}Zr_{12}V_{13}Cu_4Be_9$, $Ti_{58}Zr_{16}V_{10}Cu_4Be_{12}$, $Ti_{46}Zr_{20}V_{12}Cu_5Be_{17}$, and $Ti_{40}Zr_{24}V_{12}Cu_5Be_{19}$ bulk metallic glass matrix composites in NaCl solution shifted to more positive values with increasing Ti and became more negative with decreasing Zr and Be. Cracks were formed around the second phase in the amorphous matrix suggesting that the amorphous matrix was more vulnerable to corrosion attack. The matrix was depleted of elements such as Ti and contained a higher fraction of Beryllium (Be) in comparison to the crystalline dendrites. This promoted local galvanic coupling followed by selective dissolution of Be and subsequent crack formation around dendrites. $Ti_{62}Zr_{12}V_{13}Cu_4Be_9$ BMGMC was more resistant to selective corrosion of the amorphous matrix due to the higher amount of Ti that formed a protective Ti-enriched oxide film [36]. Table 5.2 is a summary of the corrosion results for different Ti-based BMGs in various environments.

Table 5.2. Corrosion behavior of Ti-based BMG alloys in different environments.

Ti-Based BMG	i_{corr} (μA/cm^2)	E_{corr} (mV vs. SCE)	E_{pit} (mV vs. SCE)	Environment	T (°C)	Ref.
$Ti_{50}Cu_{28}Ni_{15}Sn_7$ (0% CNT)	18.07	−273	249	Hank's	37	[3]
$Ti_{50}Cu_{28}Ni_{15}Sn_7$ (4% CNT)	33.55	−276	-	Hank's	37	[3]
$Ti_{50}Cu_{28}Ni_{15}Sn_7$ (8% CNT)	23.95	−328	-	Hank's	37	[3]
$Ti_{50}Cu_{28}Ni_{15}Sn_7$ (12% CNT)	10.58	−276	-	Hank's	37	[3]
$Ti_{47}Cu_{38}Zr_{7.5}Fe_{2.5}Sn_2Si_1Ag_2$	0.65	−110	990	PBS	37	[37]
$Ti_{46}Cu_{27.5}Zr_{11.5}Co_7Sn_3Si_1Ag_4$	268	−270	-	PBS	37	[6]
$Ti_{46}Cu_{27.5}Zr_{11.5}Co_7Sn_3Si_1Ag_4$	201	−151	-	0.15 M NaCl	25	[6]
$Ti_{46}Cu_{27.5}Zr_{11.5}Co_7Sn_3Si_1Ag_4$	155	−289	-	1 M HCl	25	[6]
$Ti_{46}Cu_{27.5}Zr_{11.5}Co_7Sn_3Si_1Ag_4$	143	−345	-	1 M NaOH	25	[6]
$Ti_{40}Cu_{36}Pd_{14}Zr_{10}$ (thin film)	0.76	−177	-	SBF	37	[38]
$Ti_{45}(Zr-Be-Cu-Ni)_{50}Nb_5$	25.3	−259	-	1 M H_2SO_4 + 2 ppm F^-	80	[33]
$Ti_{45}(Zr-Be-Cu-Ni)_{45}Nb_{10}$	22.9	−246	-	1 M H_2SO_4 + 2 ppm F^-	80	[33]
$Ti_{45}(Zr-Be-Cu-Ni)_{40}Nb_{15}$	25.1	−266	-	1 M H_2SO_4 + 2 ppm F^-	80	[33]
$Ti_{41.3}Cu_{43.7}Hf_{13.9}Si_{1.1}$	-	−165	-	NaCl	25	[5]
$Ti_{41.3}Cu_{43.7}Hf_{13.9}Si_{1.1}$	-	−624	-	Hank's	25	[5]

Table 5.2. Cont.

Ti-Based BMG	i_{corr} ($\mu A/cm^2$)	E_{corr} (mV vs. SCE)	E_{pit} (mV vs. SCE)	Environment	T (°C)	Ref.
$Ti_{40}Zr_{29}Be_{16}Cu_8Ni_7$	5.46	−216	-	1 M H_2SO_4 + 2 ppm F^-	80	[32]
$(Ti_{40}Zr_{29}Be_{16}Cu_8Ni_7)_{99}V_1$	8.66	−127	-	1 M H_2SO_4 + 2 ppm F^-	80	[32]
$(Ti_{40}Zr_{29}Be_{16}Cu_8Ni_7)_{97}V_3$	8.11	−249	-	1 M H_2SO_4 + 2 ppm F^-	80	[32]
$(Ti_{40}Zr_{29}Be_{16}Cu_8Ni_7)_{95}V_5$	7.04	−240	-	1 M H_2SO_4 + 2 ppm F^-	80	[32]
$(Ti_{40}Zr_{29}Be_{16}Cu_8Ni_7)_{97}Ta_3$	4.26	−284	-	1 M H_2SO_4 + 2 ppm F^-	80	[32]
$(Ti_{40}Zr_{29}Be_{16}Cu_8Ni_7)_{95}Nb_5$	6.31	−223	-	1 M H_2SO_4 + 2 ppm F^-	80	[32]
$Ti_{40}Zr_{35}Cu_{17}S_8$	i_{Pass} = 7.5i	~−300	~−900	0.1 M NaCl	25	[39]
$Ti_{50}Zr_{25}Cu_{17}S_8$	i_{Pass} = 6.7	~−300	~−1000	0.1 M NaCl	25	[39]
$Ti_{40}Zr_{24}V_{12}Cu_5Be_{19}$	2.25	−328	-	10% H_2SO_4	25	[34]
$Ti_{40}Zr_{24}V_{12}Cu_5Be_{19}$	8.32	−1445	-	40% NaOH	25	[34]
$Ti_{40}Zr_{24}V_{12}Cu_5Be_{19}$	6.40	−469	-	0.6 M NaCl	25	[34]

5.3. Magnesium (Mg)-Based Metallic Glasses

Magnesium and its alloys are very attractive as biodegradable implant materials and show excellent biocompatibility, low density, and an elastic modulus close to the human bone [40]. Mg-based BMGs have recently attracted attention as biodegradable implant materials [41]. Mg-Zn-Ca BMGs with varying proportion of each element show unique and attractive attributes for biomedical applications [1,40, 42–44]. This system consists of non-toxic elements [41] and represents a great choice as a biodegradable bone tissue scaffold material [45]. The results of cytotoxicity and cell culture tests confirmed higher cell viability in Mg–Zn–Ca BMGs compared to crystalline Mg alloys [43]. The effect of each of the three components (i.e., Mg, Zn, and Ca) on corrosion behavior of the alloys was analyzed through potentiodynamic polarization experiments. Mg alloys with the lowest Ca content (i.e., 4–6 at. %) had the lowest corrosion current density and good passivation characteristics, while no passivation was observed when Ca content exceeded 50 at. %. The addition of Zn increased the charge transfer resistance of the alloy and improved its corrosion resistance [46]. The Zn-rich $Mg_{60}Zn_{35}Ca_5$ amorphous alloy exhibited a lower corrosion rate compared with $Mg_{66}Zn_{30}Ca_4$ in SBF [41]. $Mg_{66}Zn_{30}Ca_4$ showed more uniform corrosion than $Mg_{70}Zn_{25}Ca_5$ [43], indicating that higher Zn content improved the corrosion resistance. A porous crystalline Zn layer was present on the surface of low-Zn containing alloys whereas a dense Zn and oxygen-rich amorphous layer was formed for Zn-rich alloys [42]. The lower corrosion rate of the $Mg_{60}Zn_{35}Ca_5$ alloy was attributed to the presence of $CaMg_2$ and $CaZn_2$ intermetallic phases. $CaZn_2$ formed $CaZn_2(PO_4)_2 \cdot 2H_2O$ in SBF which is insoluble in the medium and enhanced the corrosion resistance of the alloy. Additionally, $CaMg_2$ was able to reduce the corrosion rate in the presence of Zn [41]. However, high Zn content was found to reduce the glass-forming ability of the Mg-Zn-Ca amorphous alloys. In electrochemistry and corrosion studies, Pourbaix diagrams map out possible stable phases for different redox states of all elements in an alloy as a function of pH. These plots can predict the stable redox species formed on the surface as a function of potential and pH in aqueous solutions. The stability regions typically consist of immunity, corrosion, and passivation, which are generally depicted with solid lines, while the water redox reactions (water stability window) are plotted as dotted lines. The Pourbaix diagram for a specific Mg-Zn-Ca alloy in Figure 5.1 shows the formation of Zn hydroxide at high potentials and pH of ~7.4. ZnO and $ZnCO_3$ surface films are formed at higher pH which reduces hydrogen gas evolution [42]. The primary passivation occurs via the formation of $Mg(OH)_2$, $Zn(OH)_2$, $Ca(OH)_2$, and ZnO_2-containing films on the surface. The presence of MgO and $Mg(OH)_2$ surface films for $Mg_{60}Cu_{30}Y_{10}$ BMG after immersion in NaCl was confirmed by X-ray photoelectron spectroscopy (XPS) analysis, although it showed low stability in the presence of Cl^- ions and dissolved gradually with the formation of $MgCl_2$ [47].

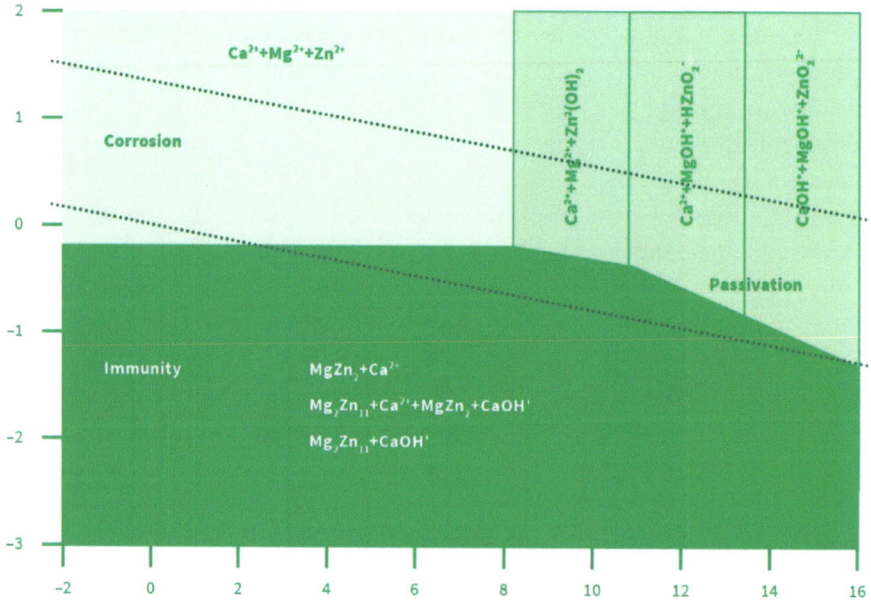

Figure 5.1. Potential–pH (Pourbaix) diagram for $Mg_{66}Zn_{30}Ca_4$ alloy (Redrawn using data from reference [41]).

5.3.1. Effect of Alloying Elements

The addition of Ti and Cr to a Mg-based BMG, specifically the $(Mg_{65}Cu_{20}Y_{10}Zn_5)_{98}M_2$ (M=Ti, Cr) alloy system, showed lower passive current density and increased corrosion potential in the Cl^--containing solution due to the formation of a more homogenous protective surface layer. However, the addition of Ti and Cr reduced the glass-forming ability of the alloy [48]. The addition of Yttrium (Y) improved the corrosion resistance of the Mg-Cu amorphous alloy system in NaCl [47]. $Mg_{65}Cu_{7.5}Ni_{7.5}Ag_5Zn_5Gd_5Y_5$ BMG in alkaline solution (NaOH) formed a passive layer consisting of $Mg(OH)_2$ with the presence of other elements such as silver (Ag), rare-earth elements (RE), and Ni [49]. The addition of ytterbium (Yb) to MgZnCa BMGs not only enhanced their ductility but also improved in vitro biocompatibility [50]. The addition of lithium (Li) to Mg-Zn-Ca BMG, specifically the $Mg_{66-x}Li_xZn_{30}Ca_4$ (x = 2, 3, 4, and 5 at. %) alloy system, showed significantly improved corrosion resistance in SBF with an increase in Li content [51]. Immersion in SBF led to the formation of $Mg(OH)_2$, LiOH, and $Ca(OH)_2$ surface passivation layers. $Mg(OH)_2$ is porous and dissolves in chloride-containing solutions [52]. However, hydrolysis of Li increased the local pH value which in turn stabilized the $Mg(OH)_2$ film, indicating potential use of this alloy system as a biomedical implant material [51]. Micro-alloying Mg-BMGs with Ni, Sr, Pd, and Ag also enhanced corrosion performance [52–56].

5.3.2. Effect of Structure and Crystallinity

There is limited understanding in terms of the electrochemical properties of Mg-based amorphous alloys compared with their crystalline counterparts. The corrosion behavior of the $Mg_{65}Y_{10}Cu_{25}$ amorphous alloy was found to be superior compared to the crystalline alloy with the same composition in borate buffer solution (pH=8.4) and NaOH solution (pH=13) due to heterogeneity and galvanic coupling in the crystalline alloy [57,58]. The same behavior was reported for $Mg_{65}Y_{10}Cu_{15}Ag_{10}$ [55]. On the other hand, partially crystallized Mg–Zn–Ca alloy showed the best corrosion performance in terms of pitting resistance, whereas the fully amorphous alloy showed the highest pitting susceptibility [59]. The $Mg_{60}Zn_{35}Ca_5$ and $Mg_{66}Zn_{30}Ca_4$ alloys showed better corrosion performance in their partially crystallized state compared to a fully crystallized state which was attributed to the formation of a more protective passive film [41].

5.3.3. Mg-Based Metallic Glass Composites

Brittle failure of Mg-based BMGs limits their widespread use. Therefore, Mg-based bulk metallic glass matrix composites (BMGMCs) with crystalline phases distributed in the amorphous matrix have been developed. $(Mg_{65}Cu_{10}Ni_{10}Y_{10}Zn_5)_{91}Zr_9$ BMGMC showed slightly lower corrosion resistance than monolithic Mg-based BMGs but significantly better than crystalline magnesium alloys, since it contained large amounts of alloying elements facilitating passive film formation [60]. In another investigation, $Mg_{69}Zn_{27}Ca_4$ BMGs reinforced with ductile Fe particles ($Mg_{69}Zn_{27}Ca_4$/Fe) spontaneously passivated with a wide passive region at a low passivation current density in NaCl solution, showing better corrosion resistance compared with AZ31 and pure Mg [61].

5.4. Calcium (Ca)-Based Bulk Metallic Glasses

Ca-Mg-Zn alloys are the most studied Ca-based BMGs due to their low density, low cost, and elastic modulus similar to the human bone [45,62]. However, these alloys are susceptible to high reactivity and fast degradation in most environments. Therefore, some form of surface modification [45] and/or compositional and structural changes [1,63,64] become necessary to make potential use of this alloy system. Ca-based amorphous alloys such as Ca-Mg-Cu, Ca-Mg-Zn, and Ca-Mg-Ag-Cu were first reported in 2002 [65–67]. $Ca_{65}Mg_{15}Zn_{20}$ exhibited rapid corrosion and spallation of corrosion products such as $Ca(OH)_2$, $Ca[Zn(OH)_3]_2 \cdot H_2O$, and Ca_3Zn. The $Ca_{50}Mg_{20}Cu_{30}$ amorphous alloy showed minor weight gain with parabolic time dependence due to oxide layer formation followed by rapid weight loss due to corrosion product spallation. However, the addition of Cu improved the corrosion resistance of this alloy system in distilled water with the formation of a three-layer protective oxide film in the $Ca_{55}Mg_{18}Zn_{11}Cu_{16}$ amorphous alloy [63]. In addition, the five-component $Ca_{55}Mg_{15}Al_{10}Zn_{15}Cu_5$ amorphous alloy demonstrated much better corrosion behavior in the aqueous environment compared to $Ca_{65}Mg_{15}Zn_{20}$, $Ca_{50}Mg_{20}Cu_{30}$, and $Ca_{55}Mg_{18}Zn_{11}Cu_{16}$, indicating that the corrosion resistance of Ca-based BMGs may be significantly improved through alloying with Zn, Cu, and Al [63,68]. The addition of Li also improved the corrosion resistance of Ca-Mg-Zn metallic glasses [69].

5.5. Aluminum (Al)-Based Bulk Metallic Glasses

Al-based amorphous alloys show outstanding mechanical properties such as high specific strength and good plasticity under compression, low density, and good corrosion resistance [70]. Attempts to develop Al-rich BMGs for engineering applications have been largely unsuccessful due to the poor glass-forming ability of these systems [71]. Recently, a new class of Al-based BMGs with Al content of 14–40 at. % was developed with the addition of rare-earth elements such as La, Ce, Gd, Y, and Er [72]. However, there are no reports on the corrosion behavior of these amorphous alloys. Despite the initial interesting findings on the corrosion resistance of Al-based glassy ribbons, such as those containing transition metals (TM) and rare-earth elements [70,73,74], there is large scope for the development of new alloys and detailed investigations on their corrosion characteristics.

References

1. Cao, J.D.; Laws, K.J.; Birbilis, N.; Ferry, M. Potentiodynamic polarisation study of bulk metallic glasses based on the Mg–Zn–Ca ternary system. *Corros. Eng. Sci. Technol.* **2012**, *47*, 329–334. [CrossRef]
2. Liu, Y.; Pang, S.; Li, H.; Hu, Q.; Chen, B.; Zhang, T. Formation and properties of Ti-based Ti-Zr-Cu-Fe-Sn-Si bulk metallic glasses with different (Ti + Zr)/Cu ratios for biomedical application. *Intermetallics* **2016**, *72*, 36–43. [CrossRef]
3. Hsu, C.-F.; Kai, W.; Lin, H.-M.; Lin, C.-K.; Lee, P.-Y. Fabrication and corrosion behavior of Ti-based bulk metallic glass composites containing carbon nanotubes. *J. Alloy. Compound.* **2010**, *504S*, S176–S179. [CrossRef]
4. Oak, J.-J.; Inoue, A. Formation, mechanical properties and corrosion resistance of Ti–Pd base glassy alloys. *J. Non-Cryst. Solids* **2008**, *354*, 1828–1832. [CrossRef]
5. Wang, G.; Fan, H.; Huang, Y.; Shen, J.; Chen, Z. A new TiCuHfSi bulk metallic glass with potential for biomedical applications. *Mater. Des.* **2014**, *54*, 251–255. [CrossRef]
6. Wang, T.; Wu, Y.D.; Si, J.J.; Cai, Y.H.; Chen, X.H.; Hui, X.D. Novel Ti-based bulk metallic glasses with superior plastic yielding strength and corrosion resistance. *Mater. Sci. Eng. A* **2015**, *642*, 297–303. [CrossRef]
7. Qin, F.; Zhu, S.; Dan, Z.; Kawashima, A.; Xie, G. Stress corrosion cracking and bioactivity of Ti-based bulk metallic glass. *J. Alloy. Compd.* **2014**, *615*, 123–127. [CrossRef]
8. Morrison, M.; Buchanan, R.; Peker, A.; Liaw, P.; Horton, J. Electrochemical behavior of a Ti-based bulk metallic glass. *J. Non-Cryst. Solids* **2007**, *353*, 2115–2124. [CrossRef]
9. Calin, M.; Gebert, A.; Ghinea, A.; Gostin, P.; Abdi, S.; Mickel, C.; Eckert, J. Designing biocompatible Ti-based metallic glasses for implant applications. *Mater. Sci. Eng. C* **2013**, *33*, 875–883. [CrossRef]
10. Oak, J.-J.; Louzguine-Luzgin, D.V.; Inoue, A. Investigation of glass-forming ability, deformation and corrosion behavior of Ni-free Ti-based BMG alloys designed for application as dental implants. *Mater. Sci. Eng. C* **2009**, *29*, 322–327. [CrossRef]
11. Oak, J.-J.; Inoue, A. Attempt to develop Ti-based amorphous alloys for biomaterials. *Mater. Sci. Eng. A* **2007**, *449–451*, 220–224. [CrossRef]
12. Bai, L.; Cui, C.; Wang, Q.; Bu, S.; Qi, Y. Ti–Zr–Fe–Si system amorphous alloys with excellent biocompatibility. *J. Non-Cryst. Solids* **2008**, *354*, 3935–3938. [CrossRef]
13. Qin, F.; Wang, X.; Zhu, S.; Kawashima, A.; Asami, K.; Inoue, A. Fabrication and Corrosion Property of Novel Ti-Based Bulk Glassy Alloys without Ni. *Mater. Trans.* **2007**, *48*, 515–518. [CrossRef]
14. Zhu, S.; Wang, X.; Qin, F.; Inoue, A. A new Ti-based bulk glassy alloy with potential for biomedical application. *Mater. Sci. Eng. A* **2007**, *459*, 233–237. [CrossRef]
15. Liens, A.; Etiemble, A.; Rivory, P.; Balvay, S.; Pelletier, J.M.; Cardinal, S.; Fabrègue, D.; Kato, H.; Steyer, P.; Munhoz, T.; et al. On the Potential of Bulk Metallic Glasses for Dental Implantology: Case Study on $Ti_{40}Zr_{10}Cu_{36}Pd_{14}$. *Materials* **2018**, *11*, 249. [CrossRef] [PubMed]

16. Qin, F.; Wang, X.; Xie, G.; Zhu, S.; Kawashima, A.; Asami, K.; Inoue, A. Microstructure and Corrosion Resistance of Ti–Zr–Cu–Pd–Sn Glassy and Nanocrystalline Alloys. *Mater. Trans.* **2007**, *48*, 167–170. [CrossRef]
17. Zhu, S.; Xie, G.; Qin, F.; Wang, X.; Inoue, A. Effect of Minor Sn Additions on the Formation and Properties of TiCuZrPd Bulk Glassy Alloy. *Mater. Trans.* **2012**, *53*, 500–503. [CrossRef]
18. Oak, J.; Louzguine-Luzgin, D.; Inoue, A. Fabrication of Ni-free Ti-based bulk-metallic glassy alloy having potential for application as biomaterial, and investigation of its mechanical properties, corrosion, and crystallization behavior. *J. Mater. Res.* **2007**, *22*, 1346–1353. [CrossRef]
19. Xie, G.; Qin, F.; Zhu, S.; Louzguine-Lugzin, D. Corrosion behaviour of porous Ni-free Ti-based bulk metallic glass produced by spark plasma sintering in Hank's solution. *Intermetallics* **2014**, *44*, 55–59. [CrossRef]
20. Xie, G.; Qin, F.; Zhu, S.; Inoue, A. Ni-free Ti-based bulk metallic glass with potential for biomedical applications produced by spark plasma sintering. *Intermetallics* **2012**, *29*, 99–103. [CrossRef]
21. Gong, P.; Deng, L.; Jin, J.; Wang, S.; Yao, K. Review on the Research and Development of Ti-Based Bulk Metallic Glasses. *Metals* **2016**, *6*, 264. [CrossRef]
22. Ke, J.L.; Huang, C.H.; Chen, Y.H.; Tsai, W.; Huang, J. In vitro biocompatibility response of Ti–Zr–Si thin film metallic glasses. *Appl. Surf. Sci.* **2014**, *322*, 41–46. [CrossRef]
23. Qin, F.; Wang, X.; Kawashima, A.; Zhu, S.; Inoue, A. Corrosion Behavior of Ti-Based Metallic Glasses. *Mater. Trans.* **2006**, *47*, 1934–1937. [CrossRef]
24. Fornell, J.; Pellicer, E.; Steenberge, N.V.; González, S.; Gebert, A.; Suriñach, S.; Baró, M.D.; Sort, J. Improved plasticity and corrosion behavior in Ti–Zr–Cu–Pd metallic glass with minor additions of Nb: An alloy composition intended for biomedical applications. *Mater. Sci. Eng. A* **2013**, *559*, 159–164. [CrossRef]
25. Abdi, S.; Oswald, S.; Gostin, P.F.; Helth, A.; Sort, J.; Baro, M.D.; Calin, M.; Schultz, L.; Gebert, A. Designing new biocompatible glass-forming $Ti_{75-x}Zr_{10}Nb_xSi_{15}$ (x= 0, 15) alloys: Corrosion, passivity, and apatite formation. *J. Biomed. Mater. Res. Part B Appl. Biomater.* **2014**, *104*, 27–38. [CrossRef]
26. Qin, F.X.; Zhou, Y.; Ji, C.; Dan, Z.H.; Yang, S. Enhanced Mechanical Properties, Corrosion Behavior and Bioactivity of Ti-based Bulk Metallic Glasses with Minor Addition Elements. *Acta Metall. Sin. (Engl. Lett.)* **2016**, *29*, 1011–1018. [CrossRef]
27. Sypien, A.; Czeppe, T. Properties of the $Ti_{40}Zr_{10}Cu_{36}Pd_{14}$ BMG Modified by Sn and Nb Additions. *J. Mater. Eng. Perform. Jmepeg* **2016**, *25*, 800–808. [CrossRef]
28. Wang, Y.; Li, H.; Cheng, Y.; Zheng, Y.; Ruan, L. In vitro and in vivo studies on Ti-based bulk metallic glass as potential dental implant material. *Mater. Sci. Eng. C* **2013**, *33*, 3489–3497. [CrossRef]
29. Qin, F.; Dan, Z.; Wang, X.; Inoue, A. Ti-based Bulk Metallic Glasses for Biomedical Applications. In *Biomedical Engineering: Trends in Materials*; Laskovski, A.N., Ed.; IntechOpen: Rijeka, Croatia, 2011; pp. 249–268.

30. Qin, F.; Yoshimura, M.; Wang, X.; Zhu, S.; Kawashima, A.; Inoue, A. Corrosion Behavior of a Ti-Based Bulk Metallic Glass and Its Crystalline Alloys. *Mater. Trans.* **2007**, *48*, 1855–1858. [CrossRef]
31. Huang, C.H.; Lai, J.J.; Wei, T.Y.; Chen, Y.H.; Wang, X.; Huang, C. Improvement of bio-corrosion resistance for Ti42Zr40Si15Ta3 metallic glasses in simulated body fluid by annealing within supercooled liquid region. *Mater. Sci. Eng. C* **2015**, *52*, 144–150. [CrossRef]
32. Debnath, M.R.; Chang, H.-J.; Fleury, E. Effect of group 5 elements on the formation and corrosion behavior of Ti-based BMG matrix composites reinforced by icosahedral quasicrystalline phase. *J. Alloy. Compd.* **2014**, *612*, 134–142. [CrossRef]
33. Debnath, M.R.; Kim, D.-H.; Fleury, E. Dependency of the corrosion properties of in-situ Ti-based BMG matrix composites with the volume fraction of crystalline phase. *Intermetallics* **2012**, *22*, 255–259. [CrossRef]
34. Yang, F.; Tian, H.; Lan, A.; Zhou, H.; Wang, B.; Yang, H.; Qiao, A.J. Corrosion Behavior of Ti-Based In Situ Dendrite-Reinforced Metallic Glass Matrix Composites in Various Solutions. *Metall. Mater. Trans. A* **2015**, *46A*, 2399–2403. [CrossRef]
35. Khalifa, H.E.; Vecchio, K.S. High Strength $(Ti_{58}Ni_{28}Cu_8Si_4Sn_2)_{100-x}Mo_x$ Nanoeutectic Matrix–β-Ti Dendrite, BMG-Derived Composites with Enhanced Plasticity and Corrosion Resistance. *Adv. Eng. Mater.* **2009**, *11*, 885–891. [CrossRef]
36. Xu, K.K.; Lan, A.D.; Yang, H.J.; Qiao, W. Corrosion behavior and pitting susceptibility of in-situ Ti-based metallic glass matrix composites in 3.5 wt.% NaCl solutions. *Appl. Surf. Sci.* **2017**, *423*, 90–99. [CrossRef]
37. Pang, S.; Liu, Y.; Li, H.; Sun, L.; Li, Y.; Zhang, T. New Ti-based Ti–Cu–Zr–Fe–Sn–Si–Ag bulk metallic glass for biomedical applications. *J. Alloy. Compd.* **2015**, *625*, 323–327. [CrossRef]
38. Subramanian, B. In vitro corrosion and biocompatibility screening of sputtered $Ti_{40}Cu_{36}Pd_{14}Zr_{10}$ thin film metallic glasses on steels. *Mater. Sci. Eng. C* **2015**, *47*, 48–56. [CrossRef]
39. Kuball, A.; Gross, O.; Bochtler, B.; Adam, B.; Ruschel, L.; Zamanzade, M.; Busch, R. Development and characterization of titanium-based bulk metallic glasses. *J. Alloy. Compd.* **2019**, *790*, 337–346. [CrossRef]
40. Li, H.; Liu, Y.; Pang, S.; Liaw, P.K.; Zhang, T. Corrosion fatigue behavior of a Mg-based bulk metallic glass in a simulated physiological environment. *Intermetallics* **2016**, *73*, 31–39. [CrossRef]
41. Ramya, M.; Sarwat, S.G.; Udhayabanu, V.; Subramanian, S.; Raj, B.; Ravi, K. Role of partially amorphous structure and alloying elements on the corrosion behavior of Mg–Zn–Ca bulk metallic glass for biomedical applications. *Mater. Des.* **2015**, *86*, 829–835. [CrossRef]
42. Zberg, B.; Uggowitzer, P.; Loffler, J. MgZnCa glasses without clinically observable hydrogen evolution for biodegradable implants. *Nat. Mater.* **2009**, *8*, 887–891. [CrossRef] [PubMed]

43. Gu, X.; Zheng, Y.; Zhong, S.; Xi, T.; Wang, W. Corrosion of, and cellular responses to Mg–Zn–Ca bulk metallic glasses. *Biomaterials* **2010**, *31*, 1093–1103. [CrossRef] [PubMed]
44. Meagher, P.; O'Cearbhaill, E.D.; Browne, J. Bulk Metallic Glasses for Implantable Medical Devices and Surgical Tools. *Adv. Mater.* **2016**, *28*, 5755–5762. [CrossRef] [PubMed]
45. Li, H.; Wang, Y.; Cheng, Y.; Zheng, Y. Surface modification of $Ca_{60}Mg_{15}Zn_{25}$ bulk metallic glass for slowing down its biodegradation rate in water solution. *Mater. Lett.* **2010**, *64*, 1462–1464. [CrossRef]
46. Zhang, S.; Zhang, X.; Zhao, C.; Li, Y.S.J.; Xie, C.; Tao, H.; Zhang, Y.; He, Y.; Jiang, Y.; Bian, Y. Research on an Mg–Zn alloy as a degradable biomaterial. *Acta Biomater.* **2010**, *6*, 626–640. [CrossRef]
47. Babilasa, R.; Bajorek, A.; Simka, W.; Babilas, D. Study on corrosion behavior of Mg-based bulk metallic glasses in NaCl solution. *Electrochim. Acta* **2016**, *209*, 632–642. [CrossRef]
48. Zhang, X.; Chen, G. The effects of microalloying on thermal stability, mechanical property and corrosion resistance of Mg-based bulk metallic glasses. *J. Non-Cryst. Solids* **2012**, *358*, 1319–1323. [CrossRef]
49. Gebert, A.; Haehnel, V.; Park, E.; Kim, D.; Schultz, L. Corrosion behaviour of $Mg_{65}Cu_{7.5}Ni_{7.5}Ag_5Zn_5Gd_5Y_5$ bulk metallic glass in aqueous environments. *Electrochim. Acta* **2008**, *53*, 3403–3411. [CrossRef]
50. Yu, H.J.; Wang, J.Q.; Dmitri, X.T.S.; Louzguine-Luzgin, V.; Perepezko, H. Ductile Biodegradable Mg-Based Metallic Glasses with Excellent Biocompatibility. *Adv. Funct. Mater.* **2013**, *23*, 4793–4800. [CrossRef]
51. Meifeng, H.; Hao, W.; Kunguang, Z.; Fang, L. Effects of Li addition on the corrosion behaviour and biocompatibility of Mg(Li)–Zn–Ca metallic glasses. *J. Mater. Sci.* **2018**, *53*, 9928–9942. [CrossRef]
52. González, S.; Pellicer, E.; Fornell, J.; Blanquer, A.; Barrios, L.; Ibáñez, E.; Solsona, P.; Suriñach, S.; Baró, M.D.; Sort, J. Improved mechanical performance and delayed corrosion phenomena in biodegradable Mg–Zn–Ca alloys through Pd-alloying. *J. Mech. Behav. Biomed. Mater.* **2012**, *6*, 53–62. [CrossRef] [PubMed]
53. Li, H.; Pang, S.; Liu, Y.; Sun, L.; Zhang, T. Biodegradable Mg–Zn–Ca–Sr bulk metallic glasses with enhanced corrosion performance for biomedical applications. *Mater. Des.* **2015**, *67*, 9–19. [CrossRef]
54. Li, H.; Pang, S.; Liu, Y.; Zhang, T. In vitro investigation of Mg–Zn–Ca–Ag bulk metallic glasses for biomedical applications. *J. Non-Cryst. Solids* **2015**, *427*, 134–138. [CrossRef]
55. Gebert, A.; Rao, R.S.; Wolffa, U.; Baunack, S.; Schultza, L. Corrosion behaviour of the $Mg_{65}Y_{10}Cu_{15}Ag_{10}$ bulk metallic glass. *Mater. Sci. Eng. A* **2004**, *375–377*, 280–284. [CrossRef]
56. Yuan, G.; Inoue, A. Mg-based bulk glassy alloys with high strength above 900 MPa and plastic strain. *J. Mater. Res.* **2005**, *20*, 394–400. [CrossRef]
57. Gebert, A.; Wolff, U.; John, A.; Schultz, L. Stability of the bulk glass-forming $Mg_{65}Y_{10}Cu_{25}$ alloy in aqueous electrolytes. *Mater. Sci. Eng. A* **2001**, *299*, 125–135. [CrossRef]
58. Gebert; Wolff, U.; Eckert, J. Corrosion Behavior of $Mg_{65}Y_{10}Cu_{25}$ Metallic Glass. *Scr. Mater.* **2000**, *43*, 279–283. [CrossRef]

59. Wang, Y.; Tan, M.J.; Pang, J.; Jarfors, E. In vitro corrosion behaviors of $Mg_{67}Zn_{28}Ca_5$ alloy: From amorphous to crystalline. *Mater. Chem. Phys.* **2012**, *134*, 1079–1087. [CrossRef]
60. Zhang, X.; Sun, J.; Luo, J.; Cheng, J. Mechanical and corrosion behaviour of in situ intermetallic phases reinforced Mg-based glass composite. *Mater. Sci. Technol.* **2017**, *33*, 1186–1191. [CrossRef]
61. Wang, J.; Huang, S.; Wei, Y.; Pana, F. Enhanced mechanical properties and corrosion resistance of a Mg–Zn–Ca bulk metallic glass composite by Fe particle addition. *Mater. Lett.* **2013**, *91*, 311–314. [CrossRef]
62. Wang, Y.B.; Xie, X.H.; Li, H.F.; Wang, X.L.; Zhao, M.Z.; Zhang, E.W.; Bai, Y.J.; Qin, L. Biodegradable CaMgZn bulk metallic glass for potential skeletal application. *Acta Biomater.* **2011**, *7*, 3196–3208. [CrossRef] [PubMed]
63. Dahlman, J.; Senkov, O.N.; Scott, J.M.; Miracle, D.B. Corrosion Properties of Ca Based Bulk Metallic Glasses. *Mater. Trans.* **2007**, *48*, 1850–1854. [CrossRef]
64. Cao, J.; Kirkland, N.; Laws, K.; Birbilis, N.; Ferry, M. Ca–Mg–Zn bulk metallic glasses as bioresorbable metals. *Acta Biomater.* **2012**, *8*, 2375–2383. [CrossRef] [PubMed]
65. Amiya, K.; Inoue, A.A. Formation and Thermal Stability of Ca-Mg-Ag-Cu Bulk Glassy Alloys. *Mater. Trans.* **2002**, *43*, 81–84. [CrossRef]
66. Senkov, O.N.; Scott, J.; Miracle, D.B. Composition Range and Glass Forming Ability of Ternary Ca-Mg-Cu Bulk Metallic Glasses. *J. Alloy. Comp.* **2006**, *424*, 394–399. [CrossRef]
67. Senkov, O.N.; Scott, J. Formation and Thermal Stability of Ca-Mg-Zn and Ca-Mg-Zn-Cu Bulk Metallic Glasses. *Mater. Lett.* **2004**, *58*, 1375–1378. [CrossRef]
68. Senkov, O.N.; Miracle, D.B.; Liaw, P.K. Development and Characterization of Low-Density Ca-Based Bulk Metallic Glasses: An Overview. *Metall. Mater. Trans. A* **2008**, *39A*, 1888–1900. [CrossRef]
69. Li, J.L.; Zhao, D.Q.; Wang, W.H. Dissoluble and degradable CaLi-based metallic glasses. *J. Non-Cryst. Solids* **2011**, *357*, 236–239. [CrossRef]
70. Li, G.; Wang, W.; Ma, H.; Li, R.; Zhang, Z.; Niu, Y.; Qu, D. Effect of different annealing atmospheres on crystallization and corrosion resistance of $Al_{86}Ni_9La_5$ amorphous alloy. *Mater. Chem. Phys.* **2011**, *125*, 136–142. [CrossRef]
71. Yang, B.; Yao, J.; Zhang, J.; Yang, H.; Wang, J.; Ma, E. Al-rich bulk metallic glasses with plasticity and ultrahigh specific strength. *Scr. Mater.* **2009**, *61*, 423–426. [CrossRef]
72. Sun, B.; Pan, M.; Zhao, D.; Wang, W.; Xi, X.; Sandor, M.; Wu, Y. Aluminum-rich bulk metallic glasses. *Scr. Mater.* **2008**, *59*, 1159–1162. [CrossRef]
73. Sweitzer, J.; Shiflet, G.; Scully, J.R. Localized corrosion of $Al_{90}Fe_5Gd_5$ and $Al_{87}Ni_{8.7}Y_{4.3}$ alloys in the amorphous, nanocrystalline and crystalline states: Resistance to micrometer-scale pit formation. *Electrochim. Acta* **2003**, *48*, 1223–1234. [CrossRef]
74. Jakab, M.A.; Scully, J.R. On-demand release of corrosion-inhibiting ions from amorphous Al–Co–Ce alloys. *Nat. Mater.* **2005**, *4*, 667–670. [CrossRef] [PubMed]

6. Noble Metal- and Rare-Earth-Based Metallic Glasses

6.1. Noble Metal-Based BMGs

A number of noble metal-based bulk metallic glasses (BMGs) have been successfully developed over the years, including Au-, Ag-, Pd-, and Pt-based alloys. Many of these compositions have good glass-forming ability, thermal stability, high strength, and good corrosion resistance [1–5]. In particular, some of the Pd-based BMGs demonstrate excellent glass-forming ability [6] and can be produced in bulk form with diameters up to 76 mm [7]. These alloys showed high activity in a number of catalytic reactions due to uniform distribution of unsaturated active sites on their surface [8]. Their catalytic activity has been correlated with work function, which is a measure of surface charge transfer and may influence the corrosion rate. Thermoplastic processing has been utilized for certain Pd-based BMGs for applications in microelectromechanical systems (MEMS) [9], electrocatalysis [10], and biomedical implants [11]. Recently, a quaternary $Pd_{79}Au_{1.5}Ag_3Si_{16.5}$ BMG with critical casting thickness of 3 mm was developed with corrosion resistance better than SUS316L in HCl and H_2SO_4 attributed to stable surface passivation layer [6]. Bulk glass forming $Pd_{43}Ni_{10}Cu_{27}P_{20}$ alloy exhibited lower corrosion rates than 316L SS, Ti-6Al-4V, and Co-Cr-Mo alloys in HCl and was virtually unaffected during immersion tests in saline (NaCl) or strong alkaline (NaOH) electrolytes [12]. Since Cu and Ni are more electro-active than Pd, their selective dissolution occurs faster when the alloy is immersed in an appropriate electrolyte leading to a nano-porous Pd-P surface layer. This selective leaching or dealloying process has been utilized in Zr-based BMGs [13], Pt-based BMGs [10], as well as Ni-based alloys [14] for electrocatalysis and other surface functionalization applications.

A comparison of corrosion behavior for amorphous and crystalline $Pd_{40}Ni_{40}P_{20}$ in different electrolytes showed lower corrosion current density (lower corrosion rate) for the crystallized state in all electrolytes studied (NaCl, HCl, H_2SO_4, and HNO_3) as compared to the fully amorphous state [7]. This was attributed to the formation of inert and corrosion-resistant phosphides on the alloy surface. Similar results on corrosion behavior were reported for some Fe-based BMGs as well [15,16]. On the other hand, crystallized $Pd_{48.2}Fe_{17}Co_{16.7}Si_{13.4}B_{4.7}$ and $Pd_{51.4}Fe_{18}Co_{18}Si_{11.1}B_{1.5}$ alloys in de-aerated buffer solution showed higher corrosion rate compared to the cast glassy alloys due to high defect density in the passive film of the annealed samples and galvanic coupling [17].

6.2. Rare-Earth Elements-Based BMGs

Recently, numerous investigations have been conducted on rare-earth (RE)-based amorphous alloys due to their good glass-forming ability, low glass transition temperature (T_g), and excellent physical and chemical properties including hydrogen storage characteristics, high thermal stability, magnetic/magneto-optical properties, corrosion resistance, and excellent mechanical properties [18,19]. La-based [20], Nd-based [21], Pr-based [22], Ce-based [23], and Gd-based [24] BMGs have been developed successfully so far with good thermal stability and mechanical properties. However, fewer investigations have focused on their chemical and electrochemical properties such as corrosion, reactivity, and oxidation behavior.

Corrosion studies for Pr-Cu-Ni-Al BMGs in NaCl solution [25] showed lack of passivation in the Cl^--ion environment. The dissolution rates in anodic over-potentials were high for all the samples. Furthermore, the alloys with higher GFA demonstrated higher corrosion resistance and the compositions with higher Cu content showed lower oxidation resistance. For $Pr_{60}Fe_{30-x}TM_xAl_{10}$ (TM = Mn, Ni, Cu, Ti; x = 0, 5 at. %) BMG in NaCl solution [26], the corrosion resistance increased, moving from $Pr_{60}Fe_{25}Mn_5Al_{10}$ to $Pr_{60}Fe_{25}Ti_5Al_{10}$, while the Mn-containing alloy showed the highest GFA.

Strontium (Sr)-based BMGs are attractive for biodegradable implant applications [27]. They exhibit low modulus, low density, low T_g, and good chemical stability [28]. Some Sr-based BMGs such as $Sr_{60}Mg_{18}Zn_{22}$ and $Sr_{60}Li_5Mg_{15}Zn_{20}$ showed fast degradation rates in deionized water [28] indicating limited use. However, certain compositions such as $Sr_{40}Mg_{20}Zn_{15}Yb_{20}Cu_5$ showed good stability [27] and better corrosion resistance compared to $Ca_{60}Mg_{15}Zn_{25}$ BMG, conventional Mg, and Mg alloys in Hank's solution with satisfactory cell viability and biocompatibility. This was attributed to the presence of Yb and Cu. Yb-containing BMG has also been reported to be chemically stable in deionized water with a reasonable degradation rate in Hank's solution [29].

Gd-based BMGs represent another RE-based amorphous system with unique magnetic and magneto-optical properties [30–32]. High temperature annealing has been shown to improve the corrosion behavior of $RE_{65}Co_{25}Al_{10}$ (RE=Gd, Ce, La, Pr, and Sm) BMGs [33] attributed to annihilation of excess free volume and formation of a dense surface oxide layer. Pre-compression at room temperature has also been shown to improve the corrosion behavior of Gd-based BMGs [34]. In addition to annealing, superheating treatment before casting has been shown to influence the corrosion behavior of $Gd_{55}Al_{25}Cu_{10}Co_{10}$ BMG [32]. Increasing the superheat temperature reduced the degree of local ordering and improved the glass-forming ability (GFA) as well as corrosion resistance. XPS analysis confirmed the presence of corrosion-resistant Gd-, Al-, Cu-, and Co-oxides after immersion in NaCl solution. Additionally, re-melting treatment improved the corrosion behavior of $Gd_{56}Al_{26}Co_{18}$ and $Sm_{56}Al_{26}Co_{18}$ amorphous alloys due to free volume reduction and homogenization [35]. The Yb-Ca-Zn-Mg system is another recently developed RE-based BMG in which the presence of Yb and Ca shifted the corrosion potential to nobler values and lowered corrosion current density in Na_2SO_4 [36]. So far, many new BMGs have been developed with minor addition of RE elements including Sc, La, Ce, Nd, Sm, Er, Ho, Tb, Gd, Yb, and Lu resulting in good mechanical and magnetic properties [37].

References

1. Chen, Y.; Chu, J.; Jang, J.; Wu, C. Thermoplastic deformation and micro/nano-replication of an Au-based bulk metallic glass in the supercooled liquid region. *Mater. Sci. Eng. A* **2012**, *556*, 488–493. [CrossRef]
2. Tang, T.; Chang, Y.; Huang, J.; Gao, Q.; Jang, J.; Tsao, C.Y. On thermomechanical properties of Au–Ag–Pd–Cu–Si bulk metallic glass. *Mater. Chem. Phys.* **2009**, *116*, 569–572. [CrossRef]
3. Laws, K.J.; Shamlaye, K.F.; Ferry, M. Synthesis of Ag-based bulk metallic glass in the Ag–Mg–Ca–[Cu] alloy system. *J. Alloy. Compd.* **2012**, *513*, 10–13. [CrossRef]
4. Legg, B.A.; Schroers, J.; Busch, R. Thermodynamics, kinetics, and crystallization of $Pt_{57.3}Cu_{14.6}Ni_{5.3}P_{22.8}$ bulk metallic glass. *Acta Mater.* **2007**, *55*, 1109–1116. [CrossRef]
5. Mozgovoy, S.; Heinrich, J.; Klotz, U.; Busch, R. Investigation of mechanical, corrosion and optical properties of an 18 carat Au-Cu-Si-Ag-Pd bulk metallic glass. *Intermetallics* **2010**, *18*, 2289–2291. [CrossRef]
6. Chen, N.; Qin, C.; Xi, G.; Louzguine-Luzgin, D.; Inoue, A. Investigation of a ductile and corrosion-resistant $Pd_{79}Au_{1.5}Ag_3Si_{16.5}$ bulk metallic glass. *J. Mater. Res.* **2010**, *25*, 1943–1949. [CrossRef]
7. Wu, Y.; Chiang, W.C.; Chu, J.; Nieh, T.G.; Kawamura, Y.; Wu, J.K. Corrosion resistance of amorphous and crystalline $Pd_{40}Ni_{40}P_{20}$ alloys in aqueous solutions. *Mater. Lett.* **2006**, *60*, 2416–2418. [CrossRef]
8. Hasannaeimi, V.; Mukherjee, S. Noble-Metal based Metallic Glasses as Highly Catalytic Materials for Hydrogen Oxidation Reaction in Fuel Cells. *Sci. Rep.* **2019**, *9*, 12136. [CrossRef]
9. Kumar, G.; Tang, H.; Schroers, J. Nanomoulding with amorphous metals. *Nature* **2009**, *457*, 868. [CrossRef]
10. Doubek, G.; Sekol, R.C.; Li, J.; Ryu, W.H.; Gittleson, F.S.; Nejati, S.; Moy, E.; Reid, C.; Carmo, M.; Linardi, M.; et al. Guided Evolution of Bulk Metallic Glass Nanostructures: A Platform for Designing 3D Electrocatalytic Surfaces. *Adv. Mater.* **2016**, *28*, 1940–1949. [CrossRef]
11. Ratner, B.; Hoffman, A.; Schoen, F.; Lemons, J. *Biomaterials Science: An Introduction to Materials in Medicine*, 2nd ed.; Academic Press: San Diego, CA, USA, 2004.

12. Watanabe, L.; Roberts, S.; Baca, N.; Wiest, A.; Garrett, S.; Conner, R. Fatigue and corrosion of a Pd-based bulk metallic glass in various environments. *Mater. Sci. Eng. C* **2013**, *33*, 4021–4025. [CrossRef] [PubMed]
13. Wiest, G.; Wang, L.; Huang, S.; Roberts, M.D.; Demetriou, P.K.; Johnson, W.L. Corrosion and corrosion fatigue of Vitreloy glasses containing low fractions of late transition metals. *Scr. Mater.* **2010**, *62*, 540–543. [CrossRef]
14. Mukherjee, S.; Sekol, R.C.; Carmo, M.; Altman, E.I.; Taylor, A.D.; Schroers, J. Tunable Hierarchical Metallic-Glass Nanostructures. *Adv. Funct. Mater.* **2013**, *23*, 2708–2713. [CrossRef]
15. Szewieczek, D.; Tyrlik-Held, J.; Paszenda, Z. Corrosion investigations of nanocrystalline iron based alloy. *J. Mater. Process. Technol.* **1998**, *78*, 171. [CrossRef]
16. Thorpe, S.; Ramaswami, B.; Aust, K. Corrosion and Auger Studies of a Nickel-Base Metal-Metalloid Glass. *J. Electrochem. Soc.* **1988**, *135*, 2162. [CrossRef]
17. Jang, H.J.; Kim, H.G. Effects of Crystallization on the Corrosion and Passivity of Amorphous Pd-Fe-Co-Si-B Alloys. *J. Nanomater.* **2017**, *2017*, 1–7. [CrossRef]
18. Liu, Y.; Han, S.; Hu, L.; Liu, B.; Zhao, X.; Jia, Y. Phase structure and hydrogen storage properties of REMg$_{8.35}$Ni$_{2.18}$Al$_{0.21}$ (RE = La, Ce, Pr, and Nd) hydrogen storage alloys. *J. Rare Earth* **2013**, *31*, 784e789.
19. Zheng, Z.; Wang, J.; Xing, Q.; Sun, Z.; Wang, Y. Effects of La addition on glass formation, crystallization and corrosion behavior of Gd–Al-based alloys. *J. Non-Cryst. Solids* **2013**, *379*, 54–59. [CrossRef]
20. Tang, M.; Bai, H.; Pan, M.; Zhao, D.; Wang, W. Bulk metallic superconductive La$_{60}$Cu$_{20}$Ni$_{10}$Al$_{10}$ glass. *J. Non-Cryst. Solids* **2005**, *351*, 2572–2575. [CrossRef]
21. Wei, B.; Pan, M.; Wang, W.; Han, B.; Zhang, Z.; Hu, W. Domain structure of a Nd$_{60}$Al$_{10}$Fe$_{20}$Co$_{10}$ bulk metallic glass. *Phys. Rev. B* **2001**, *64*, 012406-1-4. [CrossRef]
22. Zhao, Z.; Wen, P.; Wang, R.; Zhang, D.; Pan, X.; Wang, W. Formation and properties of Pr-based bulk metallic glasses. *J. Mater. Res.* **2006**, *21*, 369–374. [CrossRef]
23. Bian, Z.; Inoue, A. Ultra-Low Glass Transition Temperatures in Ce-Based Bulk Metallic Glasses. *Mater. Trans. Jim* **2005**, *46*, 1857–1860. [CrossRef]
24. Zhong, X.C.; Min, J.X.; Liu, Z.W.; Zheng, Z.G.; Zeng, D.C.; Franco, V. Low hysteresis and large room temperature magnetocaloric effect of Gd$_5$Si$_{2.05-x}$Ge$_{1.95-x}$Ni$_{2x}$ (2x = 0.08, 0.1) alloys. *J. Appl. Phys.* **2013**, *113*, A916. [CrossRef]
25. Meng, Q.; Zhang, S.; Li, J. Corrosion and oxidation behavior of Pr-based bulk metallic glasses. *J. Alloy. Compd.* **2008**, *452*, 273–278. [CrossRef]
26. Meng, Q.; Li, J.; Bian, X. Corrosion behavior of bulk metallic glass Pr$_{60}$Fe$_{30-x}$TM$_x$Al$_{10}$ in NaCl aqueous solution. *J. Alloy. Compd.* **2006**, *424*, 350–355. [CrossRef]
27. Li, H.F.; Zhao, K.; Wang, Y.B.; Zheng, Y.F.; Wang, W. Study on bio-corrosion and cytotoxicity of a Sr-based bulk metallic glass as potential biodegradable metal. *J. Biomed. Mater. Res. B Appl. Biomater.* **2012**, *100B*, 368–377. [CrossRef]
28. Zhao, K.; Li, J.; Zhao, D.; Pan, M.; Wang, W. Degradable Sr-based bulk metallic glasses. *Scr. Mater.* **2009**, *61*, 1091–1094. [CrossRef]

29. Zhao, K.; Jiao, W.; Ma, J.; Gao, X.Q.; Wang, W.H. Formation and properties of strontium-based bulk metallic glasses with ultralow glass transition temperature. *J. Mat. Res.* **2012**, *27*, 2593–2600. [CrossRef]
30. Ma, Y.; Yu, P.; Xia, L. Achieving the best glass former in a binary Gd–Co alloy system. *Mater. Des.* **2015**, *85*, 715–718. [CrossRef]
31. Chen, D.; Takeuchi, A.; Inoue, A. Gd–Co–Al and Gd–Ni–Al bulk metallic glasses with high glass forming ability and good mechanical properties. *Mater. Sci. Eng. A* **2007**, *457*, 226–230. [CrossRef]
32. Zhang, X.; Guo, J.; Liu, H.; Song, Y.; Xu, L.; Liu, J. Influence of melt superheat treatment on corrosion resistance of Gd-based BMG in 3.5% NaCl solution. *Mater. Des.* **2016**, *100*, 217–222. [CrossRef]
33. Zheng, Z.; Xing, Q.; Sun, Z.; Xu, J.; Zhao, Z.; Chen, S. Effects of annealing on the microstructure, corrosion resistance, and mechanical properties of $RE_{65}Co_{25}Al_{10}$ (RE = Ce, La,Pr,Sm,and Gd) bulk metallic glasses. *Mater. Sci. Eng. A* **2015**, *626*, 467–473. [CrossRef]
34. Xing, Q.; Zhang, K.; Wang, Y.; Leng, J.; Jia, H.; Liaw, P.K. Effects of pre-compression on the microstructure, mechanical properties and corrosion resistance of $RE_{65}Co_{25}Al_{10}$ (RE = Ce, La, Pr, Sm and Gd) bulk metallic glasses. *Intermetallics* **2015**, *67*, 94–101. [CrossRef]
35. Liu, J.; Hou, J.; Zhang, X.; Guo, J.; Fan, G. Influence of Remelting Treatment on Corrosion Behavior of Amorphous Alloys. *Rare Met. Mater. Eng.* **2017**, *46*, 296–300.
36. Wang, J.; Qin, J.; Gu, X.; Zheng, Y.; Bai, H. Bulk metallic glasses based on ytterbium and calcium. *J. Non-Cryst. Solids* **2011**, *357*, 1232–1234.
37. Luo, Q.; Wang, W. Rare earth based bulk metallic glasses. *J. Non-Cryst. Solids* **2009**, *355*, 759–775. [CrossRef]

7. Concluding Remarks

Numerous bulk metallic glasses (BMGs) with excellent mechanical and electrochemical properties have been developed and investigated so far. Among the known bulk glass formers, Fe-based, Ti-based, and Zr-based BMGs have shown great potential for biomedical applications with the corrosion resistance of certain compositions being better than conventional biomedical alloys such as stainless steel and Ti-6Al-4V. Owing to their homogenous structure free from microstructural defects common in polycrystalline materials, metallic glasses generally show high resistance to corrosion and pitting. Protective surface layers have typically been reported for metallic glasses containing passive film forming elements such as Cr, Mo, and Co. Although recent research has significantly advanced understanding on the corrosion behavior of amorphous alloys in different environments, there are several open questions and key issues that require further investigations. Some of the key remaining questions are listed below.

1. Long-term corrosion studies for BMGs in different environments including electrolytes mimicking the physiological environment are missing. This is critical in evaluating their worthiness in structural applications, including their use as biomedical implant materials.

2. There are very limited studies on stress corrosion cracking (SCC) and the fatigue behavior of bulk metallic glasses in different corrosive environments. More systematic studies are needed for comprehensive understanding of corrosion behavior during mechanical loading of BMGs.

3. There are very limited studies on the corrosion behavior of Al-based BMGs, noble metal-based BMGs, and rare earth element based BMGs. Understanding the electrochemical behavior of these alloys could make them potentially attractive for various functional applications.

4. Difference in cooling rate across the cross-section of cast BMG samples may result in a different structural state of the glass as well as residual stresses. How the structural state of an amorphous alloy in terms of its free volume affects its corrosion behavior is not well understood and requires systematic studies.

5. In addition to monolithic amorphous alloys, bulk metallic glass matrix composites (BMGMCs) have attracted lot of interest in recent years. BMGMCs consist of crystalline dendrites in an amorphous matrix that results in exceptionally high fracture toughness. However, there are no systematic corrosion studies in these multi-phase alloys, which may show quite complex electrochemical signature.

Understanding the passivation mechanisms in these multi-phase systems may require advanced surface characterization techniques.

6. While there are a lot of studies related to the small-scale mechanical behavior of metallic glasses, there are very few reports on local electrochemical measurements in these systems. Measurement of the local electrochemical activity of BMGs with site-specific methods such as scanning electrochemical microscopy (SECM) and scanning kelvin probe (SKP) may provide valuable insights into the mechanism for corrosion initiation in these amorphous alloy systems.

7. There are very few studies on simulation and modeling of corrosion behavior in amorphous alloys. Molecular dynamics studies may help in fundamental understanding in terms of the effect of chemistry and structure on the electrochemical behavior of BMGs.

MDPI
St. Alban-Anlage 66
4052 Basel
Switzerland
Tel. +41 61 683 77 34
Fax +41 61 302 89 18
www.mdpi.com

MDPI Books Editorial Office
E-mail: books@mdpi.com
www.mdpi.com/books